The Minerals, Metals & Materials Series

T. S. Srivatsan · Yuzheng Zhang
W. C. Harrigan, Jr.
Editors

Metal-Matrix Composites Innovations, Advances and Applications

An SMD Symposium in Honor of William C. Harrigan, Jr.

TMS

Springer

Editors
T. S. Srivatsan
The University of Akron
Akron, OH
USA

W. C. Harrigan, Jr.
Gamma Technology
Valencia, CA
USA

Yuzheng Zhang
Gamma Alloys
Valencia, CA
USA

ISSN 2367-1181 ISSN 2367-1696 (electronic)
The Minerals, Metals & Materials Series
ISBN 978-3-319-72852-0 ISBN 978-3-319-72853-7 (eBook)
https://doi.org/10.1007/978-3-319-72853-7

Library of Congress Control Number: 2017961486

Printed on acid-free paper

This Springer imprint is published by Springer Nature
The registered company is Springer International Publishing AG
The registered company address is: Gewerbestrasse 11, 6330 Cham, Switzerland

Preface

The development, emergence, and proliferation in the use of metal-matrix composites have shown a noticeable increase in recent years, often replacing the heavier monolithic materials for use in a spectrum of performance critical and nonperformance critical applications. The high demand for these composites is also seen in the industries spanning air transportation, ground transportation, defense-related products, and even a plethora of commercial products to include sporting goods. To explore the growth of knowledge on aspects related and relevant to composite materials research, development, emergence, and use in appropriate applications and to concurrently promote communication with specific reference to innovations, advances, and applications of metal-matrix composites, this international symposium was held in honor of Dr. W. C. Harrigan, Jr., to bring to light his outstanding contributions resulting from dedicated, diligent, and dynamic contributions in the domains spanning processing, characterization, property evaluation, failure analysis, applications, and even large-scale commercialization of the family of metal-matrix composites. Appropriately, this symposium was given the title "Metal-Matrix Composites Innovations, Advances and Applications" and was held during the TMS 2018 147th Annual Meeting & Exhibition in Phoenix, Arizona, USA, March 11–15, 2018. This symposium was well represented with abstracts from the United States and international engineers and scientists from academia, research laboratories, and industries. This volume contains many of the papers presented in the five-session symposium, sponsored by the Composite Materials Committee and Structural Materials Division of The Minerals, Metals & Materials Society (TMS). Over 35 abstracts were approved for presentation (both oral and as posters) in the five sessions:

Session 1: Aluminum and Lightweight Metal Matrix Composites
Session 2: Synthesis and Developments of Emerging Composites
Session 3: Basic History and Advances in Metal Matrix Composites
Session 4: Mechanical Behavior of Metal Matrix Composites
Session 5: Poster Presentation

In the four oral sessions, recent advances in composite materials processing, characterization, properties, and modeling specifically for aerospace, marine, land applications, and commercial products were presented. Each presenter brought new knowledge and ideas to the symposium, and we, the symposium organizers, extend our warmest thanks and appreciation to the presenters and participants. We also extend our most sincere thanks and appreciation to the elected and governing representatives of the Composite Materials Committee and Structural Materials Division of TMS for their understanding and acknowledgment of our interest, and timely approval of our request to organize this intellectually stimulating symposium.

Unbounded gratitude and valued appreciation and applause, having resonating significance, are also extended to the following individuals:

(i) **Ms. Trudi Dunlap** (TMS Manager of Events, Education & Exhibitions) for her sustained attention, assistance, interest, involvement, and timely participation stemming from understanding. Ms. Dunlap ensured a timely execution of the numerous intricacies related to both orchestration and layout of this symposium from the moment following its approval through the compilation and publication of this proceedings publication.

(ii) Special thanks and much deserved appreciation is extended to **Ms. Patricia Warren** (TMS Programming and Proceedings Specialist) for her patience, understanding, and much valued and desired attention to all the specifics and intricacies, far too many to list, from conception to completion by way of compilation and presentation of this volume.

(iii) The timely publication of this bound volume would not have been possible without the cooperation of the authors and the publishing staff headed by the dedicated TMS Content Senior Manager, **Mr. Matt Baker**.

We hope that this volume will provide all the readers, to include both the experienced and the new entrant, new perspectives and directions in their research efforts in the domain of research on metal-matrix composites. The growth and demand of metal-matrix composites will only show continuing upward trend, and we do hope that by putting together this monumental symposium, we have tried to both aid and assist aerospace, ground transportation, and industries dealing with commercial products in engineering the safe development and eventual use of metal-matrix composites for potentially viable applications.

Dr. T. S. Srivatsan
Dr. Yuzheng Zhang
Dr. W. C. Harrigan, Jr.

Contents

Part I Aluminum and Lightweight Metal Matrix Composites

Aluminum Matrix Composites 1970–2017 3
William C. Harrigan

Commercial-Ready and Large-Scale Manufacturing of Light-Weight Aluminum Matrix Nanocomposites 19
Yuzheng Zhang, Miguel Verduzco, Andrew Parker, Mark Sommer, William Harrigan Jr. and Al Sommer

Development of an Electroless Plating Process for Multi-wall Carbon Nanotubes (MWCNTS) to Improve Their Dispersion and Wettability in Molten Aluminum 29
Mohammed Elsharkawi and Amal M. K. Esawi

Part II Synthesis and Developments of Emerging Composites

Study on Hot Deformation Behavior and Processing Map of 20 Volume Percent $Al_{18}B_4O_{33w}$/2024 Composites 43
Wenchen Xu, Xiangqian Zeng, Xueze Jin and Debin Shan

Development and Characterization of In-situ Aluminum–Titanium Carbide Composites Prepared by Pneumatic Powder Injection Route 59
Sheetal Gupta, Anirban Giri, Saikat Adhikari and Vivek Srivastava

Part III Basic History and Advances in Metal Matrix Composites

Microstructure and Mechanical Behavior of Cryomilled Al–Mg Composites Reinforced with Nanometric Yttria Partially Stabilized Zirconia 71
Matthew Dussing, Hanry Yang, Troy D. Topping, Enrique J. Lavernia, Kaka Ma and Julie M. Schoenung

Fatigue Crack Growth Resistance of Titanium Metal Matrix Composites 87
H. Stanley, M. Dear, T. J. A. Doel and P. Bowen

Experimental Optimization of Dry Sliding Wear Behavior of Titanium Matrix Composites Using Taguchi Methods 103
Koutarou Hattori, Shogen Hirami, Yoshiko Hasegawa, Hiroshi Izui and Yoshiki Komiya

Model-Based Damage Detection in Piezoelectric Fiber Based Composites 119
Khalid M. Shalan, Mohamed E. AbdelMeguid, Tarek M. Hatem, Hesham A. Hegazi and Yehia A. Bahei-El-Din

Part IV Mechanical Behavior of Metal Matrix Composites

Effect of Matrix Properties and Sliding Counterface on the Wear Behavior of Magnesium Alloy Metal Matrix Composites 135
S. Jayalakshmi, R. Arvind Singh and T. S. Srivatsan

Synthesis and Microstructural Development of Particulate Reinforced Metal-Matrix Composites Using the Technique of Spray Atomization and Deposition 149
T. S. Srivatsan, Yaojun Lin, Fei Chen, K. Manigandan and Enrique J. Lavernia

Magnetically Induced Cavitation for the Dispersion of Particles in Liquid Metals 183
M. Sarma, I. Grants, A. Bojarevics and G. Gerbeth

An Engineered Magnesium Alloy Nanocomposite: Mechanisms Governing Microstructural Development and Mechanical Properties 193
Sravya Tekumalla, Shikhar Bharadwaj, T. S. Srivatsan and Manoj Gupta

The Tensile Response and Fracture Behavior of a Copper-Niobium Microcomposite: Role of Surface Modification 203
Paul Arindam and T. S. Srivatsan

Fundamental Issues and Highlights of Reactive Wetting in Carbon-Based Composites 221
Khurram Iqbal and Stevens Cadet

Part V Poster Session

Influence of Graphene Nanoplatelet Reinforcements on Microstructural Development and Wear Behavior of an Aluminum Alloy Nanocomposite 233
Mohammad Alipour, Reza Eslami Farsani and Yu. A. Abuzin

Author Index 247

Subject Index 249

About the Editors

T. S. Srivatsan is a Professor of Materials Science and Engineering in the Department of Mechanical Engineering at The University of Akron. He received his graduate degrees [Master of Science in Aerospace Engineering (M.S. 1981) from Georgia Institute of Technology and Doctor of Philosophy in Mechanical Engineering (Ph.D. 1984)] from Georgia Institute of Technology. Dr. Srivatsan joined the faculty in The Department of Mechanical Engineering at The University of Akron in August 1987. Since joining, he has instructed undergraduate and graduate courses in the areas of (a) Advanced Materials and Manufacturing Processes, (b) Mechanical Behavior of Materials, (c) Fatigue of Engineering Materials and Structures, (d) Fracture Mechanics, (e) Introduction to Materials Science and Engineering, (f) Mechanical Measurements, (g) Design of Mechanical Systems, and (h) Mechanical Engineering Laboratory. His research areas currently span the fatigue and fracture behavior of advanced materials to include monolithic(s), intermetallic, nano-materials, and metal-matrix composites; processing techniques for advanced materials and nanostructure materials; the interrelationship between processing and mechanical behavior; electron microscopy; failure analysis; and mechanical design. He has authored/edited/coedited fifty-nine (59) books and four (4) monographs in areas cross-pollinating mechanical design; processing and fabrication of advanced materials; deformation, fatigue, and fracture of ordered intermetallic materials; machining of composites; failure analysis; and technology of rapid solidification processing of materials. He serves as coeditor of *International Journal on Materials and Manufacturing Processes* and on the editorial advisory board of five journals in the domain of materials science and engineering. His research has enabled him to deliver over two hundred and twenty (220+) technical presentations in national and international meetings and symposia. He has authored and coauthored over seven-hundred (700+) archival publications in international journals, book chapters, proceedings of national and international conferences, book reviews, and technical reports. In recognition of his efforts and contributions and their impact on furthering science, technology, and education, he has been elected

(a) Fellow of the American Society for Materials, International (ASM International);
(b) Fellow of American Society of Mechanical Engineers (ASME); and
(c) Fellow of American Association for the Advancement of Science (AAAS).

He has also been recognized as (i) Outstanding Young Alumnus of Georgia Institute of Technology, (ii) Outstanding Research Faculty of the College of Engineering [The University of Akron], and (iii) Outstanding Research Faculty at The University of Akron. He offers his knowledge in research services to the National Research Laboratories and industries related to aerospace, automotive, power generation, leisure-related products, and applied medical sciences.

Yuzheng Zhang is Lead Materials Scientist at Gamma Alloys, LLC. Other current positions include Visiting Scholar at the USC Composites Center.

Dr. Zhang received his Ph.D. in Materials Science from the University of Southern California (USC) in 2015, M.S. in Materials Science from USC in 2010, and B.S. in Physics from Fudan University (China) in 2008.

Prior to joining Gamma, Dr. Zhang worked as a Research Scientist at Nextgen Aeronautics, developing flexible strain sensor arrays. At Gamma, he continues his interest in processing and characterizing aluminum-based MMCs, and studying fundamental strengthening mechanisms in nanocomposites.

William C. Harrigan, Jr. received a Ph.D. in Materials Science from Stanford University. He received a Master of Science from Stanford University and a Bachelor of Science in Metallurgical Engineering, Cum Laude, from the University of Notre Dame. Dr. Harrigan is the CTO of Gamma Technology which is a commercial manufacturer of metal matrix composites. Dr. Harrigan worked at the Aerospace Corporation where he developed graphite fiber reinforced aluminum and other metal composites. He worked at DWA Aluminum Composites where he helped to develop hardware made from graphite/metal composites and particle reinforced aluminum composites. Dr. Harrigan helped three companies qualify an aluminum matrix composite for use in nuclear spent fuel storage: Alyn Corporation, DWA Technologies, and Ceradyne, Inc. He also managed programs to commercialize metal matrix composite automobile engine components and recreation products. He has also consulted with several companies that are developing new manufacturing methods for aluminum matrix composites. Dr. Harrigan has taught classes in design with composites at the University of California, Irvine, and California State University, Northridge.

In 1985, Dr. Harrigan was the chairman of the International Conference on Composite Materials V, held in San Diego California. He was honored as a Fellow of ASM International in 1999 with the citation for commercialization of metal matrix composites.

While working at The Aerospace Corporation, he was awarded US Patent 4,223,075 for alloy modification for graphite fiber reinforced aluminum composites. At Alyn Corporation, he received two patents for use of metal matrix composites for computer disk drive substrates, 5,948,495 and 5,895,696.

Part I
Aluminum and Lightweight Metal Matrix Composites

Aluminum Matrix Composites 1970–2017

William C. Harrigan

Abstract Aluminum matrix composites have evolved from continuous fiber composites, boron fiber/aluminum and graphite fiber aluminum, to whisker and particle reinforced aluminum during this time. The limited applications for continuous fiber composites, Space shuttle tube truss and Hubble telescope high gain antenna boom have evolved into more wide-spread applications such as heat sinks for high power electronics, helicopter transmission parts and power transfer components, as well as spent fuel containment components for the nuclear power industry. This paper discusses the manufacturing process that have been developed to allow the production of components as well as the property combinations that made the applications viable. Single ply vacuum hot pressed composite sheets have evolved into powder consolidated 400 kg billets that are extruded and rolled or forged into final shape. All of these innovations have made the aluminum composites more attractive from a cost standpoint. This is the ultimate driving force, good properties at an acceptable price.

Keywords Aluminum matrix composites · Manufacturing · Particle reinforced aluminum

Introduction

Aluminum matrix composites have taken many forms in the past 40 years. In the 1970s the composites were continuous fiber reinforced aluminum. By the 1980s the reinforcements were whiskers and particles. The 1990s saw the development of larger structures made from particle reinforced aluminum matrix composites. After 2000 the emphasis has been on new reinforcements and nano structured matrix alloys and on nano reinforced composites.

W. C. Harrigan (✉)
Gamma Technology, Valencia, CA, USA
e-mail: Bill.Harrigan@gammaalloys.com

T. S. Srivatsan et al. (eds.), *Metal-Matrix Composites Innovations, Advances and Applications*, The Minerals, Metals & Materials Series,
https://doi.org/10.1007/978-3-319-72853-7_1

1970

The first large scale application for the continuous fiber composites was as tubes for the truss structure of the Space Shuttle, Fig. 1 [1]. The reason for this is the extremely high compression strength of the boron/Aluminum composites, 2750 MPa coupled with the low density for the composite [2]. The tubes were one third the weight of high steel tubes, with the same compression strength. Therefore, the weight savings generated by the tubes justified the cost. This composite was made by wrapping the 0.14 mm boron fibers on to a 2-m diameter drum with aluminum foil on the surface, Fig. 2. The fiber was held in place with a fugitive binder, glue, that would be off-gassed during the vacuum hot pressing operation. Once the fiber was wrapped and the glue dried, the monolayer of aluminum foil and fiber was removed from the drum by cutting the layer. The monolayers were stacked with additional aluminum foil as a single ply or multiple plies and hot pressed to produce the composite. The monolayers were then assembled as tubes, shapes or cross plies and hot pressed again to form the final product.

Another large structure that was made from continuous fiber composite is the high gain antenna booms for the Hubble Space Telescope, Fig. 3 [3]. This structure was made from graphite fiber-aluminum composite and served two purposes, it is the hardware that positions the high gain antenna for transmitting information to and from earth and it is the waveguide to carry the signal from the antenna to the

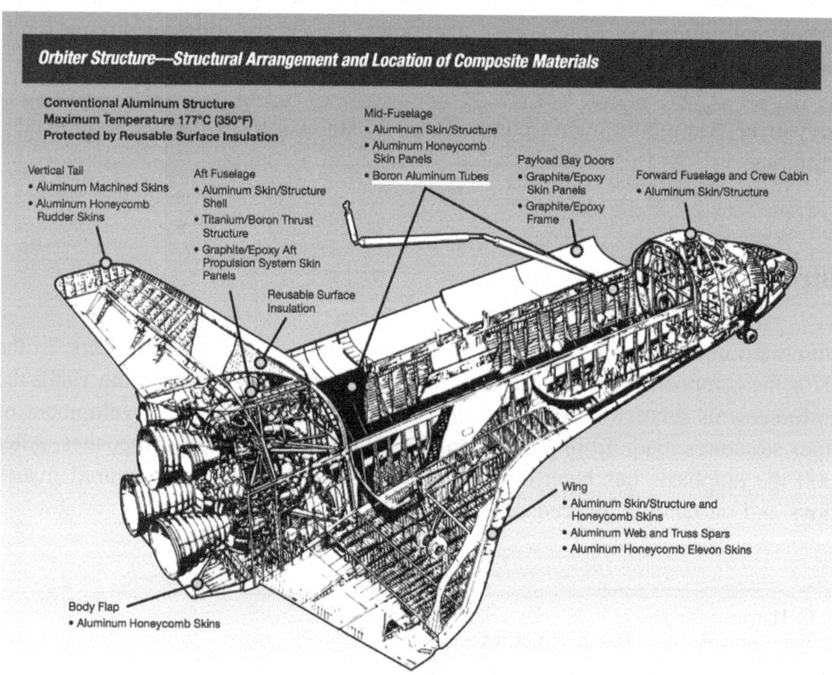

Fig. 1 Diagram of the space shuttle orbiter with components identified

Fig. 2 Photograph of wrapping drum with aluminum foil and boron fibers placed on the foil

spacecraft electronics. This aluminum-graphite structure has a unique property, a thermal expansion coefficient of zero in the length direction, resulting in high pointing accuracy for the antenna. One side of the wave guide would be 100 °C when it faced the sun, while the other side would be −150 °C as it faced away from the sun. Without the zero-expansion coefficient, the 3-m long boom would experience a warp and poor pointing accuracy would result. The inside surface of the tube is aluminum and has high electrical conductivity so the rectangular tube is a wave guide. The aluminum composite structure replaced three structures, a graphite/epoxy tube with an aluminum wave guide plus thermal insulation to protect the graphite epoxy. The weight reduction brought about by the single structure justified the use of the expensive aluminum/graphite composite.

The aluminum-graphite composite was made with a chemical vapor deposition process followed by a liquid metal infiltration process shown in Fig. 4 [4]. Tows of continuous graphite fibers were pull through a heated chamber where $TiCl_4$ and BCl_3 gasses were reduced to boron and titanium by zinc vapor. Titanium and boron were deposited on the graphite fiber surface. The coated fibers were immediately immersed in molten aluminum and were spontaneously infiltrated by the aluminum. The resulting composite was a wire containing 10,000 fibers and was 1.3 mm in diameter, Fig. 5 [5]. These wires were then vacuum hot pressed into plate structures. This process was also used to make graphite composites with matrices of magnesium, copper, lead and silver.

Fig. 3 Diagram of the hubble space telescope, showing location of aluminum/graphite boom

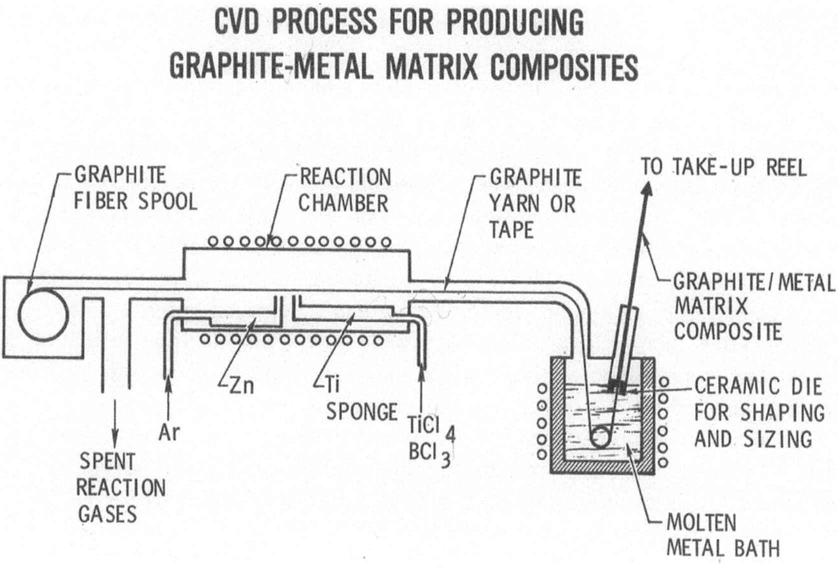

Fig. 4 Diagram of process for manufacture of graphite-metal composite

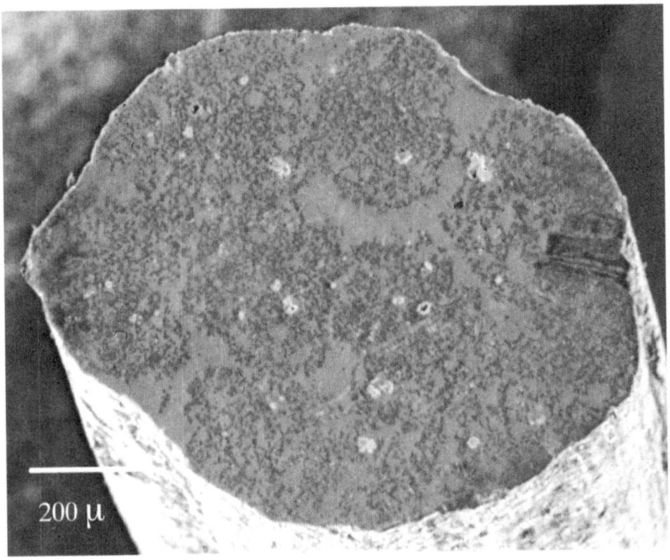

Fig. 5 SEM micrograph of a graphite fiber reinforced aluminum composite wire

Both the boron-aluminum tube struts and the graphite aluminum boom structures were used because of the tremendous premium for weight savings in space structures. With the advent of more powerful rockets, the premium for weight reduction is less and these structures would no longer be affordable.

1980

In the early 1980s a potentially low-cost manufacturing process for making silicon carbide whiskers was developed using rice hulls as the source for the silicon. These whiskers, approximately 0.5 micron in diameter and 10 microns long, were incorporated into aluminum using powder metallurgy techniques. These composites had interesting properties but suffered from the high cost of the whiskers and the health concerns for fine particles with long length to diameter, l/d, ratios, similar to asbestos. The powder metallurgy processing was extended to incorporation of low cost silicon carbide particles from the grinding and polishing industry, to make a composite with a large cost savings, Fig. 6 [6]. These composites could be extruded, rolled and forged into shapes. We no longer had to mold the composite into a final form. We were able to take advantage of existing manufacturing facilities developed for processing other metals for scale-up and to reduce the cost of our composites. The composites were isotropic, especially the particle reinforced composites and the composites could be machined without cutting fibers and compromising the performance of the structure.

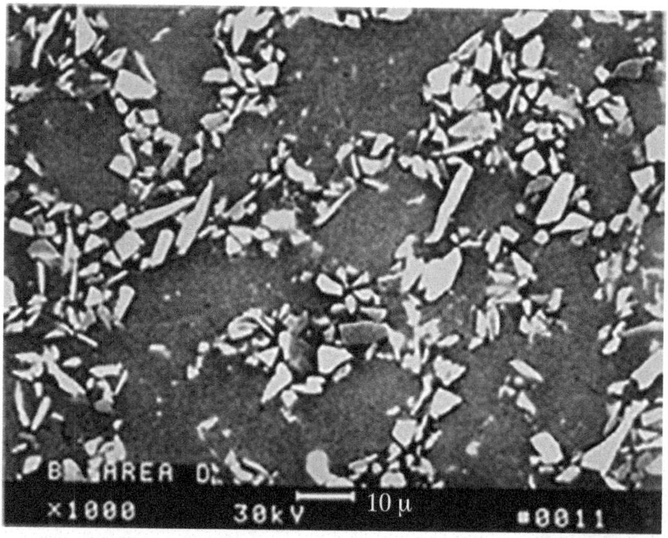

Fig. 6 SEM micrograph of silicon carbide particle reinforced aluminum composite

The mechanical properties were not as high as the continuous fiber composites in the fiber direction, but the isotropic properties, especially the high elastic modulus led to a program to replace the equipment racks in the U-2 Reconnaissance Aircraft [7]. The racks were designed for isotropic bending stiffness so high isotropic stiffness is very important. The composite elastic properties and the bending stiffness, EI calculations for three materials considered in the program are contained in Table 1 [8]. The three materials are aluminum, the baseline, quasi-isotropic graphite epoxy and a silicon carbide particle reinforced 6061 aluminum-composite. Shapes of the resulting parts are shown in Fig. 7. The area of the resulting extrusions, or pultrusion, are contained in the Fig. 7. The program demonstrated that silicon-carbide-aluminum composite could be reliably extruded with a wall thickness of 0.2 mm as required by the design.

The high stiffness of the MMC resulted in a smaller cross section and this gave rise to a lower weight for the MMC even though the density of the MMC was higher than the aluminum. The weight savings for the graphite/epoxy compared with the aluminum is 19.2%. The weight savings for the MMC compared with the aluminum is 37.4%. The weight savings for the MMC compared with the graphite/

Table 1 Design data for electronic rack extrusion

Material	Elastic modulus (GPa)	Moment of inertia (cm^4)	E*I (10^6 kg cm^2)
Aluminum	70	0.7767	0.5472
Graphite/Epoxy	57.8	0.9403	0.5472
SiC/Aluminum	107.5	0.4961	0.5468

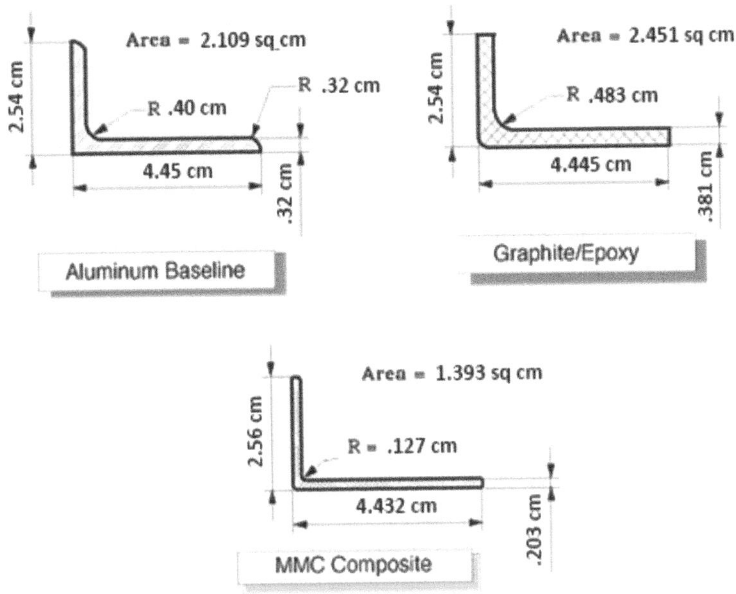

Fig. 7 Extrusion shape for electronic racks applications

Table 2 Weight calculations for rack extrusions

Material	Density (g/cc)	Area (cc)	Weight/cm (grams)	Weight/m (grams)
Aluminum	2.839	2.109	5.989	598.9
Graphite/Epoxy	1.976	2.451	4.844	484.4
MMC	2.847	1.394	3.967	396.7

epoxy is 22.6%. Therefore, the high isotropic modulus of the MMC resulted in large weight savings in this application (Table 2).

The extrusions were made from 150 mm diameter billets. In order to extrude with an area ratio of approximately 30:1, the extrusion dies needed to have two, four or six holes to produce two, four or six parts per extrusion. The use of multi-hole dies is common for extruding aluminum alloys. Over 10,000 m of extrusions were made for this program. These extrusions were solution treated as 7.5–10-m long sections in a vertical furnace. The parts were quenched into a pit and then moved to the stretch racks. A stretch of 2% was found to produce extrusion within straightness tolerance. The 7.5 m extrusions were stretched 15 cm and the 10-m section were stretched 21 cm. The pieces were then aged and delivered to the customer for rack manufacture.

As part of the material qualification, tensile tests were conducted on the lead section of every extrusion. For example, the four-hole die required 4 tensile samples

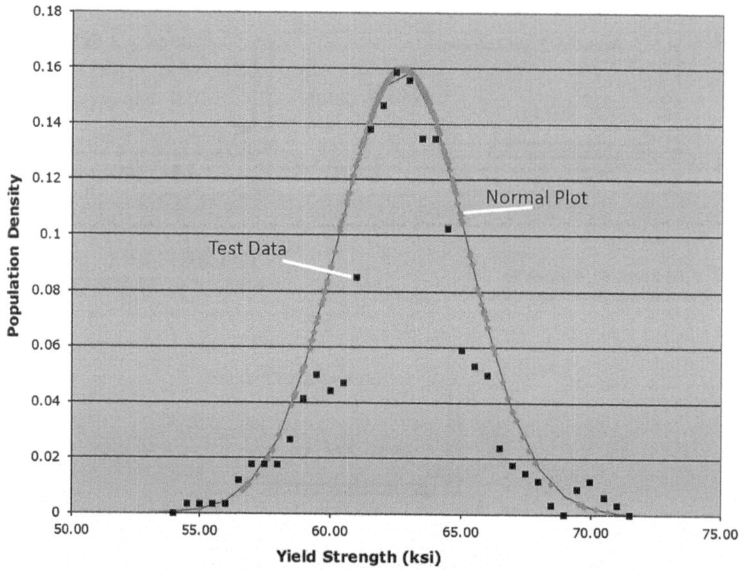

Fig. 8 Tensile test yield strength distribution for 6092/SiC/25p composite extrusions

per extruded billet. This requirement was relaxed after the initial extrusion runs and one test per billet extrusion would qualify the material. This requirement resulted in 273 samples with a single composite composition that were tensile tested. The yield strength of the material was found to be 432 MPa with a standard deviation of 17.2 MPa. A plot of the distribution of the yield strength, Fig. 8, shows that the yield strength data of these samples have a normal distribution of values. The composite could be tested using metal based standards and has properties that resemble metals.

Another project that was generated by the increased isotropic stiffness of the particle reinforced composites involved the manufacture of cross beams for the America's Cup catamaran Stars and Stripes '88, Fig. 9 [7]. The hulls of the boat were held together by two 300 mm diameter aluminum cross beams that were 9.1-m long. The cross beams were to be replaced by a particle reinforced aluminum containing 20 volume percent silicon carbide. This composite has an isotropic modulus that is 20% higher than aluminum, so the cross beam could be 20% lighter than the baseline aluminum beams. In light wind conditions, the boat weight is very important, so saving 20% of the cross-beam weight was attractive. These cross beams were made by back extruding 432 mm diameter billets 432 mm OD by 318 mm ID extrusion preforms, Fig. 10. Because of time restraints, the whole program took 10 weeks to complete, existing tube extrusion tooling was used to make the cross-beam tubes. This resulted in tubes that were identical to the original tubes. The weight savings were generated by chemical-milling the outside surface

Fig. 9 Photograph of catamaran stars and stripes ’88

Fig. 10 Photograph of 430 mm diameter MMC billet with 320 mm diameter punch and 430 mm diameter centering ring

Fig. 11 Cross beam for "stars and stripes '88" being chemical milled to reduce wall thickness

of the tubes to reduce the wall thickness from 6.35 mm to 5.1 mm. This produced tubes that were 20% lighter than the original aluminum tubes.

Locations where bolt holes were to be drilled remained 6.35 mm thick by masking these areas, Fig. 11. One of the cross beams was tested by the "Stars and Stripes '88" staff in San Diego. The tube was loaded in flexure to demonstrate the 20% increase in stiffness. A central load of 5450 kg was applied to the tube to demonstrate capability of resisting the full mast load at the center of the tube during racing conditions.

1990

By the mid-1990s the particle reinforced aluminum composites had found uses as skins for fighter aircraft, rolled sheets containing boron carbide for neutron absorbing in spent fuel canisters and many other applications. Extrusion of 400 kg billets was demonstrated. Rolling of sheets with a size of 1 m by 5 m by thickness between 2 and 30 mm was demonstrated [9]. The impressive mechanical properties, including the fatigue resistance of the particle reinforced aluminum composites were not enough to generate widespread usage. The poor machinability of the composite caused by the dulling of tools by the imbedded SiC or B4C particles drove up the cost of making structures and limited the applications for the composites.

2000

In the early 2000s, a spheroidized alumina reinforcement was developed [10]. The lower cost of the alumina particles as a reinforcement was always attractive, however, alumina particles could not use as a reinforcement for high strength aluminum matrix composites because the magnesium in the alloys would react with the alumina to form a spinel compound. The loss of the magnesium from the alloy reduces the heat treat response of the matrix alloy, resulting in low strength composites. During the spherodization process, tabular alumina is transformed to spherical alumina by melting the alumina particles. Surface tension transforms the molten alumina droplets into spheres. The rapid freezing of the small particles produces a particle that no longer reacts with magnesium in aluminum alloys and allows us to make high strength aluminum matrix composites. The sphere does not have the sharp corners present on the SiC and B4C particles used for other metal matrix composites. This lack of sharp corners reduces the tool wear significantly and allows the new composite to be machined with standard tools rather than the polycrystalline diamond tools needed for the older composites. This reduces the cost of machining the composite and expanded the potential applications for MMCs.

2010

A Navy Small Business Innovative Research Program allowed Gamma Technologies to manufacture Bearing liners for the CH 53K helicopter transmission case, Fig. 12 [11]. Bearing liners are buffers that are placed between bearings and the cast aluminum or magnesium transmission cases, Fig. 13. They protect the expensive transmission case casting from vibrational wear from the bearings. The composite used for this program is a 2000 series aluminum alloy that is reinforced with 20 volume percent spheroidized alumina. The composite is made by powder metallurgy vacuum hot pressing. The liners are machined from ring preforms. The liner preforms are made by forward extruding tubes for the smaller liners or by back-extruding tubes followed by ring rolling for the larger liners. The 20-volume percent reinforcement increased the wear resistance of the aluminum matrix to the level of case hardened steel, the liner material that was replaced. The bearing liners for the CH 53K helicopter are rings that have outside diameters ranging from 64 mm to 470 mm. The length of the liners is from 25 to 75 mm and the wall thickness of the liners is approximately 3 mm. The composite density is one third that of the steel leading to a weight reduction when a steel liner is replaced by a MMC liner with one third the weight. The aluminum MMC liners will reduce the weight of the target transmission by 25 kg. The commercial processing of the composite in coupled with the ease of machining result in bearing liners that are competitive in price with steel liners at one third the weight.

Fig. 12 Photograph of CH 35K helicopter [13]

A recent innovation is the introduction of nano size alumina particles into aluminum using a modified powder metallurgy process. The process for making these composites centers around de-agglomerating nano size alumina particles and allowing these particles to decorate the surface of aluminum powder particles. The evolution of the microstructure of these composites is shown in Fig. 14. In this figure, clean surfaces of an aluminum powder particles are decorated with alumina nano particles. The nano particles remain on the surface of the original powder particles in the as-pressed composite. After extrusion or other metal working operation, the nano particles are incorporated into the aluminum matrix, creating a nano reinforced aluminum composite.

The nano particles not only add stiffness, increase wear resistance and decrease expansion coefficient, but significantly increase the strength of the composites. The nano particles act as artificial precipitates that do not dissolve into the matrix as the temperature is raised, so the elevated temperature strength is also significantly increased. The tensile properties of a composite made from unalloyed aluminum, 1100, reinforced with nano size alumina particles, are plotted in Fig. 15. The samples were full annealed, 450 °C for 2 h plus furnace cool to 150 °C. Also

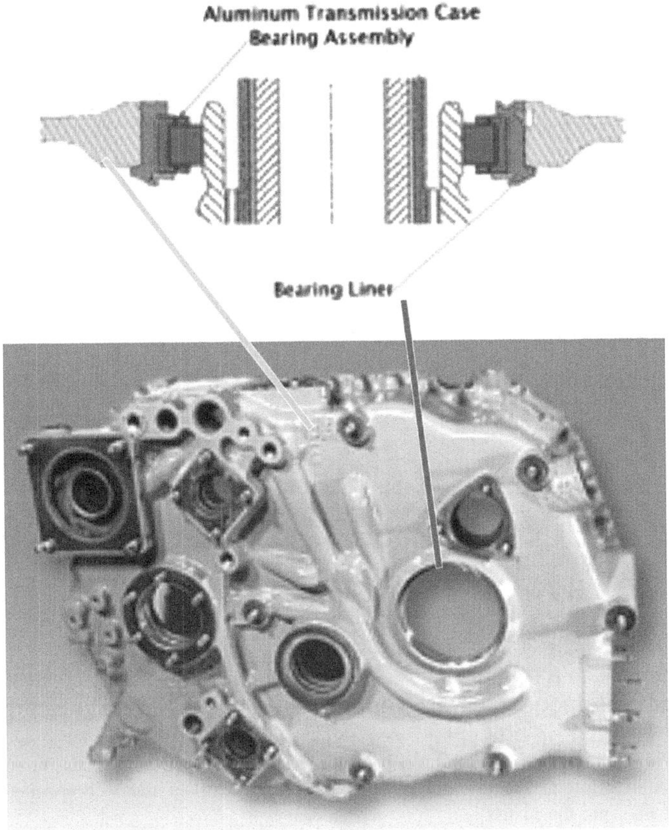

Fig. 13 Drawing of a bearing liner and Photograph of bearing liners [11]

contained in this plot are data for annealed 1100 aluminum [12] and 2618 aluminum [12] at 375 °C. The nano particles in the pure aluminum increase the room temperature yield strength from 30 MPa to 260 MPa with an eight percent addition. The same composite has a yield strength of 140 MPa at 375 °C while the 1100 aluminum only has a yield strength of 10 MPa. One of the best aluminum alloy for high temperature applications, 2618, only has a yield strength of 24 MPa [13]. The nano particles make the pure aluminum composite a "super aluminum" at elevated temperatures.

The processes that are used to make these composites have been developed around powder metallurgy procedures that can be scaled up to large scale manufacturing. The blending operation is the key to making the process work and it employs commercial processes that allow us to decorate the alumina nano particles on aluminum powder. The powder consolidation can be done by standard powder metallurgy processes, such as vacuum hot pressing, CIP/Sinter, SPS consolidation

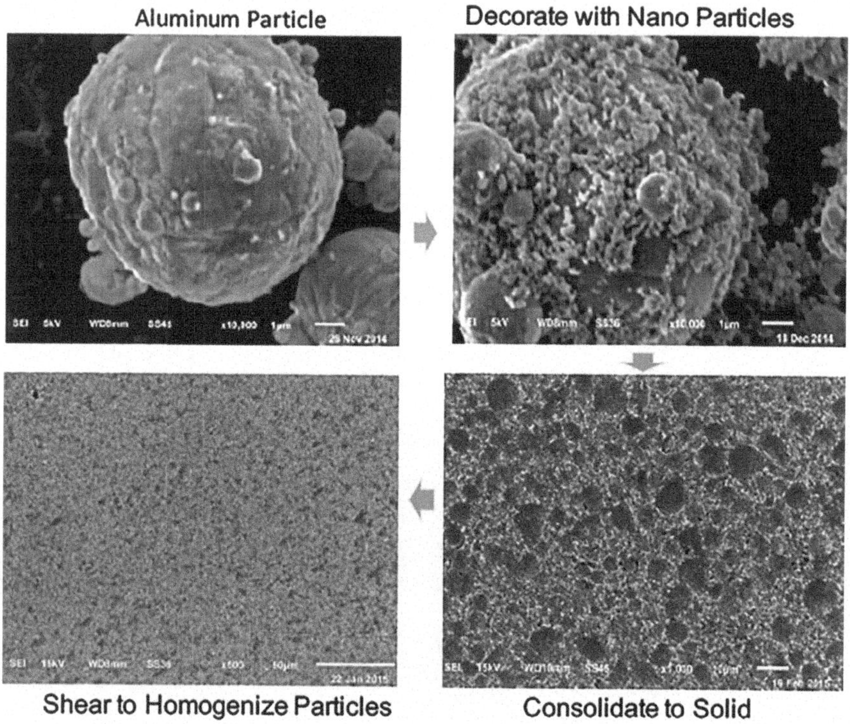

Fig. 14 SEM micrographs of aluminum particles, nano particle decorated aluminum particles, as-pressed nano reinforced aluminum composite and hot worked nano composite with nano particles sheared into the matrix

Fig. 15 Yield strength as a function of test temperature for 1100 aluminum with nano size alumina reinforcement

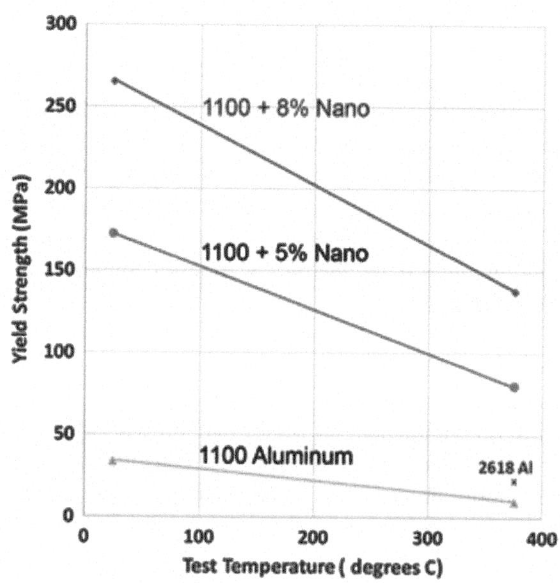

or powder forging. The metal working processes such as hot extrusion, hot forging and hot rolling that have been developed for particle reinforced aluminum composites over the years, have been used to produce nano particle reinforced aluminum composites. The raw materials for the composite are commercially available products, atomized aluminum and nano size alumina.

Summary

Aluminum matrix composites have been produced with a wide variety of reinforcements and matrix alloys. Structures with amazing properties such as the tube struts for the Space Shuttle with compressive strengths of 2750 MPa that were recently retired with the Shuttle fleet. The zero-expansion high gain antenna booms for the Hubble Space Telescope are still in operation and working perfectly. These structures were made because of a significant weight savings was realized because of unique properties of the composites. The tremendous premium on weight savings for these spacecraft is no longer present because of larger rockets and such expensive material are no longer needed and can no longer be used.

The particle reinforced aluminum composites have lower properties but their combination of properties couple with their lower cost have allowed these composites to be used in many applications. The increase in isotropic stiffness for the particle reinforced composites have enable them to be used as electronic racks in reconnaissance aircraft and as cross beams for racing catamarans. The increase fatigue resistance allowed sheet products of the composite to be used for fuel access doors on F-16 fighter aircraft. However, the limitation in machining these composites has limited their widespread use in structures. The advent of spherical alumina as a reinforcement has eliminated the poor machining aspect of the aluminum MMCs but been slow to be adopted because of prior experiences with the older composite materials.

The nano reinforcement offers high strength as well as high modulus to the property enhancements offered by the aluminum MMCs. This new composite offers property combinations that have not been possible before, especially at elevated temperatures. We need to continue to develop low cost approaches to manufacture this material while maintain the quality of the material. We are all looking for our "Golf Club Shaft" to generate enough usage of a material to reduce the cost by making large quantities. In 1982 Aldila [14] started supplying golf club shafts to equipment manufactures in quantities that allowed graphite fiber manufacturers to reduce the price of the fibers. All of the DoD efforts to incorporate the graphite fibers into fighter aircraft did not generate a large enough usage of the material to significantly increase the production volume and reduce the cost of the material. A golf club shaft did the trick!

References

1. Buck ME, Suplinskas RJ (1987) Continuous boron fiber MMCs. In: Engineered materials handbook, vol 1 Composites. ASM International, pp 851–866
2. Knoell AC (1975) Mechanical properties of boron-aluminum composite. In: AIAA Conference Paper, May 1975, pp 75–789
3. Harrigan WC (1987) Discontinuous SiC fiber MMC. In: Engineered materials handbook, vol 1 Composites. ASM International, 1987, pp 889–895
4. Harrigan WC, Jr, Flowers RH (1977) Graphite-metal composites, titanium-boron vapor deposit method of manufacture. In: Proceedings failure modes in composites, TMS-AIME Chicago, 24 Oct 1977, pp 319–335
5. Harrigan WC, Jr (1974) Fabrication of graphite-aluminum composite. Report SAMSO-TR-74-253. The Aerospace Corp, 1974.
6. Harrigan WC, Jr, Gaebler G, Davis E, Levin EJ (1982) The effects of hot-rolling on the mechanical properties of SiC reinforced 6061 aluminum. In: Proceedings mechanical behavior of metal matrix composites, 111th AIME Annual Meeting, Dallas TX, Feb 1982, pp 169–180
7. Harrigan WC, Jr (1991) Metal Matrix composites. In: Everett R (ed) Metal matrix composites processing and interfaces. Elsevier Inc., ISBN 978-0-12-341832-1
8. Harrigan WC, Jr (Unpublished research)
9. Foltz JV, Harrigan WC (1991) Powder metallurgy metal matrix composites overview. In: 1991 P/M Aerospace and Defense Technologies, published by Powder Metallurgy Industries Federation, Princeton, NJ, pp 123–128
10. Harrigan WC, Jr (2002) New reinforcements for aluminum matrix composites. In: Proceedings of the 2002 world congress on powder metallurgy and particulate materials, vol 2, Metal Powder Industries Federation, Dec 2002, pp 114–125
11. Small Business Innovative Research Program (2012) Navair contract N68335-11-0052, Final Report,Large Diameter Bearing Liners, Feb 2012
12. Kaufmann JG (1999) Properties of aluminum alloys tensile, creep and fatigue data at high and lowtemperatures. ASM & Aluminum Association 1999, ISBN 0-87170-623-6
13. Harrigan WC (2013) Presentation large diameter bearing liners. At U.S. Navy Transition Assistance Meeting, Washington DC, 17 July 2013
14. History: Aldila.com

Commercial-Ready and Large-Scale Manufacturing of Light-Weight Aluminum Matrix Nanocomposites

Yuzheng Zhang, Miguel Verduzco, Andrew Parker, Mark Sommer, William Harrigan Jr. and Al Sommer

Abstract Light-weight aluminum matrix nanocomposites (AMnCs) exhibit superior strength and stiffness compared to its alloyed counterpart, especially at elevated temperatures, which makes this material an ideal candidate for high temperature applications in automobile, aerospace and energy industries. However, the use of aluminum MMC is limited due to its prohibitively high cost in scale-up production. In 2008, a path to commercialize aluminum MMC was laid out by Gamma Alloys LLC. A unique powder metallurgy approach was used to synthesize alumina nanoparticle reinforced aluminum MMC in a cost-effective way. A uniform distribution of spherical alumina nanoparticles was achieved in the matrix using a patented powder mixing technique. This technique is compatible with all aluminum alloy systems and thus customizable based on customer's needs. Today, Gamma Alloys' materials have found numerous applications from structural parts of automotive engines, to military helicopter transmissions and cleats for athletic running shoes.

Introduction

In the past 20 years, aluminum matrix nanocomposites (AMnCs) have emerged as advanced materials which exhibit higher stiffness, improved wear resistance, excellent strength-to-weight ratio and better high temperature performance compared to the microcomposite and monolithic counterparts. Typical reinforcements in aluminum matrix include ceramic particles such as Al_2O_3, SiC, AlN, Si_3N4, TiO_2 and TiB_2, either in particular or fiber form. The key factors that define the resulting properties of AMnCs are

- Reinforcing particle size and morphology
- Particle distribution

Y. Zhang (✉) · M. Verduzco · A. Parker · M. Sommer · W. Harrigan Jr. · A. Sommer
Gamma Alloys LLC, 28128 Livingston Ave, Valencia, CA 91355, USA
e-mail: yuzheng.zhang@gammaalloys.com

T. S. Srivatsan et al. (eds.), *Metal-Matrix Composites Innovations, Advances and Applications*, The Minerals, Metals & Materials Series,
https://doi.org/10.1007/978-3-319-72853-7_2

- Particle-matrix interface
- Particle volume fraction

Many synthesis methods have been proposed and investigated to achieve desired reinforcement dispersion and good properties. Common nanocomposite fabrication methods are mechanical milling [1], stir or ultrasonic-assisted casting [2, 3], pressure infiltration [4] and in situ fabrication method [5, 6], etc. However, all existing fabrication techniques showed little or no progress on commercialization because of the technical hurdles listed below:

- Severe particle agglomeration
- Limited ingot size
- Trapped contamination
- Prohibitively high cost
- Unavailability of high quality reinforcing particles

Recently, Gamma Alloys LLC has discovered a method suitable for making high quality AMnCs in an industrial scale ($\sim$100 kg) without the limitations mentioned above. In addition, this proprietary technique is compatible with conventional metal working processes such as extrusion, forging and rolling to form desired shapes for various applications in automotive, aerospace and energy industries. Sources for high quality aluminum powders and ceramic nanoparticles have become readily available in today's market thanks to an increasing development and usage of aluminum powders in solar cells and automotive paint, and mature applications of ceramic nanoparticles in semiconductor and cosmetic industries. In this article, we will briefly describe our fabrication strategy and present the resulting microstructure as well as material properties.

Commercial-Ready Manufacturing of Aluminum Matrix Nanocomposites

Figure 1 summarizes the processing route starting from raw materials to final products with a variety of forms. After raw material inspection, matrix powders and reinforcing particles were mixed together properly using a proprietary technique at Gamma Alloys LLC. Following the powder mixing, the powders were vacuum hot pressed into a billet whose size is only limited by load capacity of the press. The consolidated billet was then extruded with a minimum extrusion ratio of 20:1 to achieve a large shearing of the matrix. Finally, post-processing turned ingots of nano composites into final products. As shown in Fig. 1, complex shapes can be achieved by extrusion, forging, machining or ring rolling. Machinability of AMnCs proved to be relatively good thanks to the presence of nanoparticles serving as chip breaker. This fabrication strategy is cost-effective and can be scaled up easily because (1) nanoparticle distribution is mainly controlled by powder mixing prior to consolidation and (2) post-processing steps involve only conventional metal working and machining.

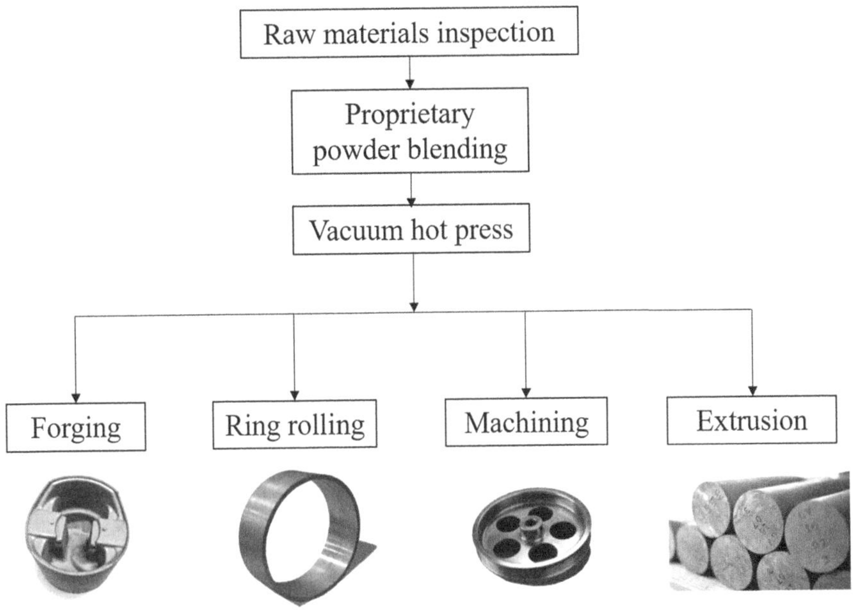

Fig. 1 Processing route of aluminum matrix nano composite

Microstructure and Strengthening Mechanisms

Figure 2 shows resulting microstructure of the aluminum matrix nanocomposite reinforced with 6 vol.% of alumina nanoparticles using a FE-SEM. As shown in Fig. 2a, a uniform and cluster-free nanoparticle distribution was achieved throughout the ingot. The matrix aluminum grains were elongated in the direction of extrusion. The light grey particles indicated by arrows in Fig. 2b are spherical alumina nanoparticles with an average size of ~50 nm. In Fig. 2b, the nanoparticle-matrix interface was clean and defect-free which prevented particle-matrix delamination during deformation and suppressed intergranular brittle fracture. The alumina nanoparticles exerted a strong pinning force on low and high angle grain boundaries at both room temperature and elevated temperature. We observed no grain growth even after the sample was exposed at 500 °C for 5 h.

The major strengthening mechanism of aluminum matrix nanocomposite is the pinning effect of alumina nanoparticles on dislocations, often called Orowan strengthening or precipitate hardening. Unlike conventional heat treatable aluminum alloys such as 2000, 6000 and 7000 series alloys, Orowan strengthening in AMnC remains effective at elevated temperature because alumina nanoparticles acting as artificial precipitates do not dissolve to the matrix at elevated temperature. As a result, AMnCs showed superior high temperature performance that is 4–10 times stronger at 350 °C (10,000 h exposure) compared to their unreinforced counterpart.

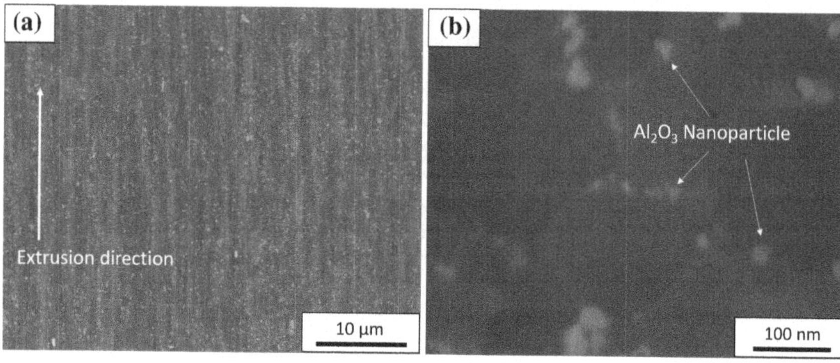

Fig. 2 Microstructure of aluminum matrix nanocomposite at: **a** 2000×, and **b** 50,000× magnification

Other strengthening mechanisms include load transfer effect, CTE mismatch between reinforcing particles and matrix, Hall-Petch strengthening and solid solution strengthening. The presence of ceramic nanoparticles strengthens the matrix via load transfer from compliant and soft matrix to stiff and hard nanoparticles. For spherical nanoparticles, the contribution of load transfer is proportional to volume fraction of the reinforcing particles and matrix yield strength [7]. A clean and defect-free interface between matrix and nanoparticles is critical to achieving an effective load transfer strengthening. Due to large CTE mismatch between alumina reinforcement (8.1×10^{-6}/°C) and aluminum matrix (23×10^{-6}/°C) [8], the formation of geometrically necessary dislocations (GNDs) at reinforcement-matrix interface strengthens the material when cooled down from the consolidation temperature (>500 °C) to room temperature. Small matrix grain size of our AMnC contributes to overall strength via Hall-Petch effect [7, 9, 10]. Because of its resistance to grain growth, our AMnC exhibits grain size strengthening effect that is insensitive to temperature. In some of our alloyed matrix, solid solution strengthening also contributes to overall strength but to a less extent compared to major mechanisms like Orowan strengthening and CTE mismatch effect.

Wear Resistance

The addition of alumina nanoparticles increased hardness of a 2000 series matrix alloy in a linear relationship. Figure 3 shows hardness in Rockwell B scale of the nanocomposite in as-consolidated condition. The horizontal dash lines represent the

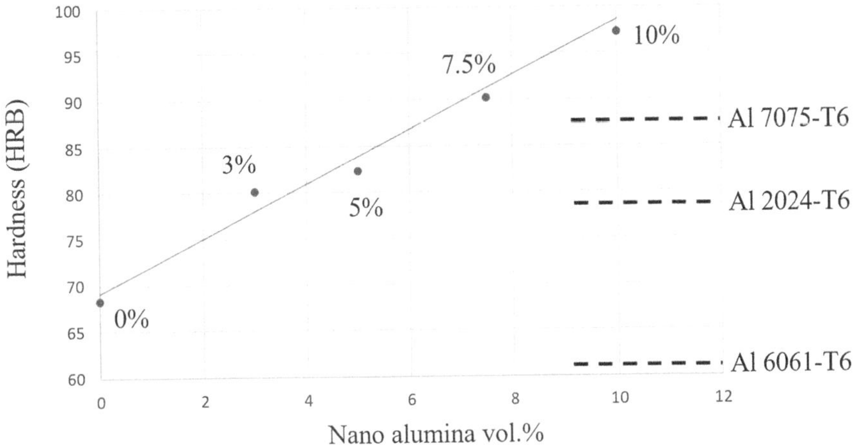

Fig. 3 Rockwell B scale hardness of a 2000 series matrix nanocomposite reinforced with 0, 3, 5, 7.5 and 10 vol.% of spherical alumina nanoparticles

hardness of commercial aluminum alloys as reference. When loaded with 7.5 vol.% of nanoparticles, AMnCs in as-consolidated condition already exhibited higher hardness than the strongest 7000 series Al-Zn alloys in T6 condition. After extrusion and T6 heat treating, a 6 vol.% 2000 series matrix nanocomposite showed a peak hardness of 107 HRB which is about 40% higher than the base alloy. Therefore, AMnCs are excellent candidates for numerous applications that require light-weight and wear-resistant materials. Wear resistance can be further enhanced using conventional hard anodizing or micro-arc oxidation.

Tensile Properties

Table 1 lists tensile properties for a 1100 Al based nanocomposite and a 2000 series nanocomposite at room temperature and elevated temperatures. The tensile properties of their unreinforced alloy counterparts were also tabulated in Table 1 as a baseline. As shown in Table 1, the tensile properties increased significantly both at room temperature and elevated temperature, while maintaining a good elongation. The yield strength of 1100 Al nanocomposite at 375 °C was 10 times higher than that of commercial 1100 Al at the same temperature. This large increase of strength at elevated temperatures was attributed to the presence of alumina nanoparticles as artificial precipitates that still providing Orowan strengthening at elevated temperatures. The nanocomposites also showed a very good strain hardenability due to a good distribution of Orowan loops at the alumina nanoparticle sites.

Table 1 Tensile properties of 1100 Al and 2000 series Al nanocomposites and their unreinforced alloy counterparts at room temperature and elevated temperatures

Temperature	Material	Yield strength (MPa)	UTS (MPa)	Elongation (%)
Room temperature	1100 Al–O [11]	34	90	40
	1100 Al + 9% Al_2O_3—full annealed	251	349	10
	2618–T6 [11]	370	440	10
	2000 series Al + 6% Al_2O_3–T6	476	579	13
Elevated temperature	1100 Al–O [11] (375 °C, after 10,000 h exposure)	11	14	85
	1100 Al + 9% Al_2O_3—full annealed (375 °C, after 10,000 h exposure)	133	135	4
	2618–T6 [11] (350 °C, after 10,000 h exposure)	28	38	95
	2000 series Al + 6% Al_2O_3–T6 (350 °C, after 10,000 h exposure)	83	93	12.5

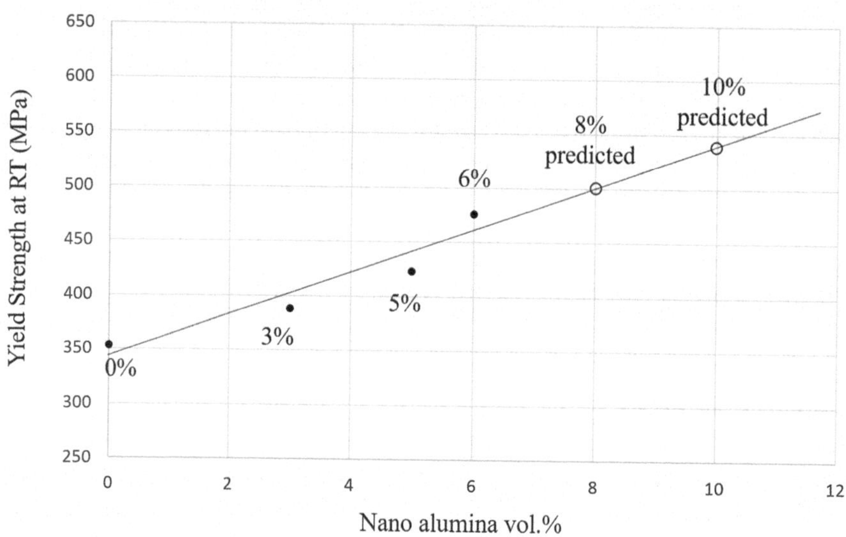

Fig. 4 Room temperature tensile yield strength of a 2000 series Al nanocomposite as a function of volume fraction of reinforcing nanoparticles

Figure 4 shows a linear relationship between room temperature yield strength and alumina nanoparticle volume fraction in a 2000 series Al nanocomposite. Eventually, this strength-nanoparticle % curve will become flat and turn downwards because nanoparticles start to agglomerate (saturation condition) as we keep increasing the volume fraction of reinforcement. We have shown that the reinforcement level can reach 12 vol.% or higher without nanoparticle agglomeration using our mixing method.

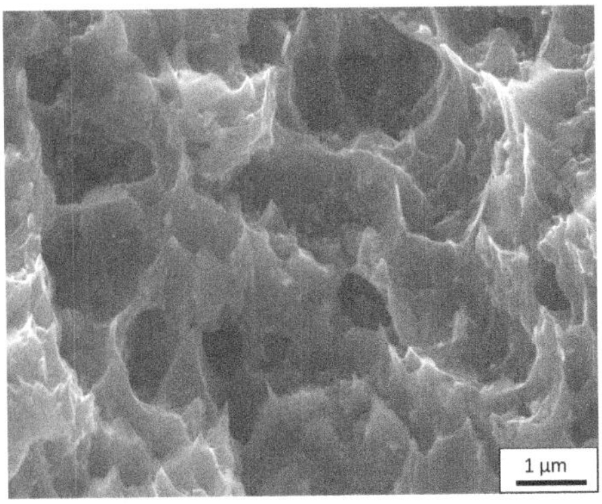

Fig. 5 Ductile fracture surface of Gamma Alloys' Al nanocomposite (6 vol.% reinforcement) with an elongation of 20% at room temperature

Many early researches reported an inferior elongation of nanocomposites due to (1) sub-ideal nanoparticle dispersion and morphology, (2) presence of micro-voids and (3) incoherent defects at nanoparticle-matrix interface. However, we found that a good elongation can be achieved in AMnCs with a cluster-free nanoparticle dispersion and clean particle-matrix interfaces. In Fig. 5, the fracture surface of the 1100 Al nanocomposite with 6 vol.% alumina nanoparticles shows a ductile fracture mode with an elongation of 20% at room temperature. No evidence of interfacial failure between nanoparticles and matrix was observed. In addition, metallographic observations using FE-SEM (Fig. 2) shows defect-free particle-metal interfaces and a clean matrix microstructure free of micro-voids.

Fatigue Resistance

To our best knowledge, no fatigue data were reported for Al nanocomposites reinforced with well dispersed ceramic nanoparticles of $\sim$50 nm. In theory, the strength of nanoparticles is related to their shear moduli. These ceramic nanoparticles should be unbreakable at the stress or stress intensity levels normally associated with aluminum matrix. Therefore, these nanoparticles will carry a majority of the structural loads applied to the composite without having particle ruptures. On the other hand, the large increases in tensile yield and ultimate strengths of nanocomposites over unreinforced alloy counterparts (as listed in Table 1) and metallographic micrographs (Figs. 2, 5) suggest a strong interface between nanoparticles and metal matrix. Based on the tensile properties and microstructures, we predict an improved fatigue resistance of our Al nanocomposites compared to conventional Al alloys. Of course, actual testing is needed to validate this prediction.

Corrosion Resistance

Thanks to its simple composition—pure aluminum matrix and chemically stable ceramic reinforcements, the 1100 Al nanocomposites are ideal candidates for those applications where corrosion resistance is a critical design factor. Corrosion resistance can be further enhanced using conventional hard anodizing or micro-arc oxidation.

Conclusions

Scale-up production and commercialization of AMnC is an important breakthrough for this light-weight advanced material with high stiffness, improved wear and corrosion resistance, and enhanced strength at both room temperature and elevated temperatures (Table 2). It is a big technical challenge, but achievable as shown in this work, to make an industrial scale nanocomposite ingot with uniform nanoparticles dispersion throughout at a reasonable cost. The basic fabrication strategy works not only for all aluminum alloys but other common automotive and aerospace materials such as titanium and magnesium alloys. Our future work will focus on fatigue behavior of AMnC, additive manufacturing of AMnC using powder bed fusion or wire based system, and exploring new matrix/reinforcement systems to meet the fast-growing needs for advanced materials in consumer electronics automotive, automotive, aerospace, and energy industries.

Table 2 Property summary of AMnCs and common aluminum alloys

Material	Density (g/cm^3)	RT specific UTS ($MPa\ cm^3/g$)	Weight Saving* (%)	Machinability	Corrosion resistance
Pure Al	2.70	33	0	Very good	Very good
Al 2024-T6	2.78	171	80	Very good	Poor
Al 3003-O	2.73	40	17	Very good	Very good
Al 5052-O	2.68	73	54	Very good	Very good
Al 6061-T6	2.7	115	71	Very good	Good
Al 7075-T6	2.81	196	83	Very good	Poor
1100 Al + 9% Al_2O_3	2.80	160	79	Good	Very good
2000 series Al +10% Al_2O_3	2.90	250	87	Good	Good

*Percentage of weight reduction to perform the same engineering task, all data reference to pure aluminum

References

1. Poirier D, Drew RAL, Trudeau ML, Gauvin R (2010) Fabrication and properties of mechanically milled alumina/aluminum nanocomposites. Mater Sci Eng A 527(29–30): 7605–7614
2. Annigeri Veeresh Kumar UKGB (2017) Method of stir casting of aluminum metal matrix composites: a review. Mater Today Proc 4(2), Part A:1140–1146
3. Murthy NV, Prasad Reddy A, Selvaraj N, Rao CSP (2017) Dispersion of alumina nano particles in Al 2219 alloy by ultrasonic assisted stir casting technique. Mater Today Proc 4 (9):10113–10117
4. Nagy Š, Nosko M, Orovčík Ľ, Iždinský K, Kúdela S, Krížik P (2015) Pre-review study of the aluminum/alumina master alloy made through pressure infiltration. Mater Des (1980–2015) 66, Part A:1-6
5. Liu X, Liu Y, Huang D, Han Q, Wang X (2017) Tailoring in-situ TiB2 particulates in aluminum matrix composites. Mater Sci Eng A 705:55–61
6. Rui-song J, Wen-hu W, Guo-dong S, Zeng-qiang W (2016) Experimental investigation on machinability of in situ formed TiB2 particles reinforced Al MMCs. J Manufact Process 23:249–257
7. Sanaty-Zadeh A (2012) Comparison between current models for the strength of particulate-reinforced metal matrix nanocomposites with emphasis on consideration of Hall–Petch effect. Mater Sci Eng A 531:112–118
8. Hidnert P, Krider HS (1952) Thermal expansion of aluminum and some aluminum alloys. J Res Nat Bur Stand 48(3):209–220
9. Hall EO (1951) The deformation and ageing of mild steel: III discussion of results. Proc Phys Soc Sect B 64:747
10. Petch NJ (1953) The cleavage strength of polycrystals. J Iron Steel Inst 174:25
11. Kaufan GJ (1999) Properties of aluminum alloys. ASM International Ohio, USA

Development of an Electroless Plating Process for Multi-wall Carbon Nanotubes (MWCNTS) to Improve Their Dispersion and Wettability in Molten Aluminum

Mohammed Elsharkawi and Amal M. K. Esawi

Abstract Embedding CNTs in molten aluminum by mechanical stirring represents one approach of producing Aluminum-Carbon Nanotube (Al-CNT) composites. Molten aluminum is known for its high surface tension with CNTs resulting in extremely poor wettability. One approach to tackle this problem is done by metallizing CNTs via electroless plating. The new interfacial metallic layer on CNTs improves their dispersion and wettability in molten aluminum. One of the challenges, however, is the rapid growth of the plated layer on CNTs during microfiltration of the prepared powder resulting in aggregates of metal-coated CNTs and grown copper crystals. This results in poor dispersion of the powder in molten aluminum making the industrial upscaling of the process not viable. In this work, chemical constraints were put to control electroless plating kinetically. Instead of growing catalytic palladium particles on top of CNTs via what is known as a two-step process sensitization and activation, colloidal ready grown Palladium-tin (Pd-Sn) particles of a fixed size were used to ensure the conformity of the coat, the concentration and volume of the catalytic solution was optimized to ensure the coverage of the surface area of CNTs. After the surface of CNTs was covered with the catalyst, a Copper-Cobalt (Cu-Co) electrolyte of extremely high deposition rate was used to ensure a reaction stopping mechanism before filtering the Copper coated CNTs ensuring that copper does not grow on aggregates of the tubes. Copper coated CNTs of an average size of 85 nm were obtained. Percentages from 0.5 to 2% of the prepared powder (containing less than 0.1% of CNTs for each 2 gm) were dispersed in molten aluminum and casted resulting into 12.2–52.1% increase in the Vickers hardness indicating the effectiveness of the copper coated CNTs in improving the properties of commercial pure aluminum.

M. Elsharkawi (✉) · A. M. K. Esawi
Department of Mechanical Engineering, The American University in Cairo (AUC), New Cairo, Egypt
e-mail: Mohammed.elsharkawi@aucegypt.edu

A. M. K. Esawi
e-mail: a.esawi@aucegypt.edu

T. S. Srivatsan et al. (eds.), *Metal-Matrix Composites Innovations, Advances and Applications*, The Minerals, Metals & Materials Series,
https://doi.org/10.1007/978-3-319-72853-7_3

Keywords Metal-matrix composites (MMCs) · Carbon nanotubes
Electroless copper-cobalt plating · Colloidal tin-palladium activation
Palladium acceleration · Casting

Introduction

Carbon nanotubes (CNTs) have been widely used for strengthening aluminum due to their high strength to weight ratio. Previous studies have pointed to the effectiveness of using CNTs in reinforcing aluminum matrices. Such studies manly reported using powder metallurgy techniques such as high energy ball milling followed by conventional and non-conventional compaction and sintering techniques to consolidate the samples [1, 2]. One of the most needed techniques in the metal industry is casting. The use of CNTs in reinforcing aluminum produced by casting has faced many stumbling blocks such as the poor wetting that limits the dispersion of CNTs in molten aluminum as well as the oxidation problem of CNTs at elevated temperatures that destroys their novel structure [3]. Recently, electroless plating has offered many advantages that allow it to work in harmony with CNTs in the casting technique. Providing a metallic coat of a controlled thickness on top of CNTs improves not only their wettability in molten metals but also, helps in their dispersion [3]. In addition, it shields the CNTs from oxidation at elevated temperatures. Generally, electroless plating has two main reactions that happen to take place simultaneously, oxidizing a reducing agent to produce electrons representing the anodic partial reaction and the reduction reaction of metal ions existing in the electroless solution by the produced electrons representing the cathodic partial reaction [4]. As the electroless plating happens to take place only on surfaces that are catalytic in nature, the surface of CNTs has to be rendered catalytic in order to be ready for electroless plating. A general scheme of catalyzing the surface is a conventional two step technique of sensitization and activation via Tin (II) chloride ($SnCl_2$) and Palladium (II) chloride ($PdCl_2$) acidic solutions [5]. This approach helped in depositing catalytic palladium particles on top of CNTs. However, the optimization process of depositing such a catalyst on CNTs is difficult due to the need of controlling four major factors represented by $SnCl_2$ concentration, $PdCl_2$ concentration, the PH level for both solutions, and the size of the produced catalytic particles. Such a process will be difficult to optimize when it comes to coating CNTs supplied from different sources because of the difference in surface area to volume ratios for different CNT's. However, electroless nickel phosphorus and copper have been reported in decorating CNTs following the previously mentioned catalyst system. Afterwards, the coated CNTs were placed in molten aluminum reporting a promising improvement in the mechanical properties of the produced composite [3]. One of the problems reported in the present work for the previous electroless approach is the absence of a reaction stopping mechanism to stop further deposition of copper on top of CNTs after the needed copper coat is achieved. This appears so important during the final filtration in the absence of sonication and stir agitation. Further depositing of copper may lead to the formation of Cu-coated CNTs aggregates. Presence of low deposition rate electrolytes makes it difficult to reach a precise reaction stopping mechanism. For electroless copper plating, the key factor of increasing the deposition rate is increasing the kinetics of the anodic partial reaction. By oxidizing more formaldehyde (reducing agent), more electrons will be produced and more copper ions will get reduced in the cathodic partial reaction. The key parameter of oxidizing the formaldehyde is the pH value that is controlled by NaOH. The higher the pH level, the higher the deposition rate. However, the pH comes to a threshold where it starts to drop when reaching a value of 12.5. Conventional copper electrolytes rely on the pH value and heating to control the kinetics of the reaction [5, 6]. In this work, a room temperature copper-cobalt electrolyte that has a high deposition rate is used. A new catalytic strategy of using colloidal palladium-tin is explored. The produced Cu-coated CNTs of percentages from 0.5–2% will be used in reinforcing aluminum using the metal casting technique. The effect of adding Cu-coated CNTs on the mechanical properties is investigated.

Materials and Experimental Procedures

Materials

Multiwalled carbon nanotubes supplied by Thomas Swan Corporation (average diameter of 10–15 nm) were used in the present study. The colloidal palladium-tin activator as well as the accelerator were supplied by Macdermid Enthony USA. Cupric SulpHate Pentahydrate (98.5% Assay) and Sodium Carbonate Anhydrous (99.5% Assay) were supplied by Lobachemie India. Sodium Hydroxide (99% Assay) was supplied by Chem-Lab Belgium. Pottasium Sodium Tartrate Tetrahydrate or what is known as Rochelle salt (99% Assay) was supplied by CARLO ERBA Reagents Italy. Cobalt (II) Chloride Hexahydrate (99% Assay) was supplied by JHD Company China. Formaldehyde 37% in aqueous solution was supplied by Alfa Aeser Germany. The pure aluminum (99.7%) was supplied by Egyptalum Company, Nagaa Hammadi, Egypt.

Experimental Procedures

The metallization of CNTs by copper was conducted in three steps. The process started by the activation of CNTs surface using colloidal Pd-Sn particles followed by the acceleration step to remove stannous hydroxide deposits on top of the

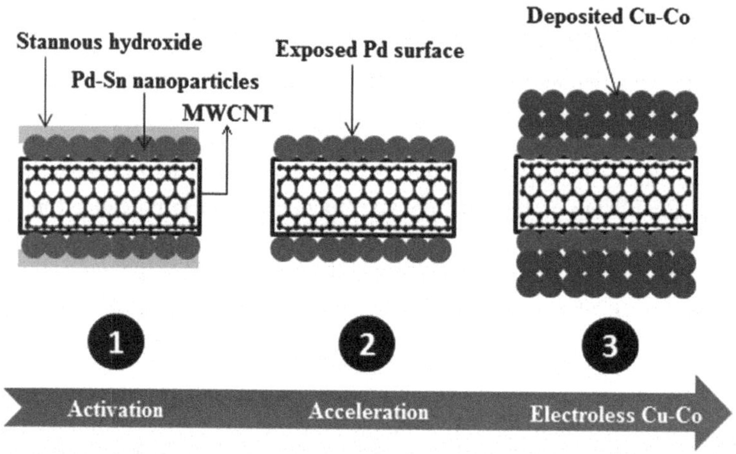

Fig. 1 General scheme of Cu-Co electroless plating on CNTs

activated surface. Finally, The electroless plating of Cu-Co on top of CNTs was performed as shown Fig. 1.

Activation of CNT's in Pd-Sn Colloidal Solution

The amount of CNTs used in this process was fixed to 0.1 gm of MWCNTs. The CNTs were used as received without any functionalization. The CNTs were activated using a commercial colloidal Pd-Sn solution adapted from the Plating on Plastics (POP) industry. These colloidal particles have an average size of 20 Angstroms and elemental ratio of 1:7 for Sn and Pd respectively [7]. The activation process was done in a mixture of DI water, HCl, and colloidal Pd-Sn solution. The different compositions listed in Table 1 were investigated to reach the optimal volumes needed for conformal electroless Cu-Co plating.

For the activation, the CNTs were dispersed by sonication in the colloidal solution for 1 min. Afterwards, the solution was put under magnetic stir agitation for 2 extra minutes. Once the stirring was finished, the treated CNTs were filtered out using 0.22 μ m PTFE filter membrane on a microfiltration kit. After filtration, CNTs were redispersed in water and refiltered again to remove the excess colloidal particles and traces of the colloidal solution from the activated CNT's. After the filtration, CNTs were collected from the membrane using tweezers.

Table 1 Typical compositions used in the optimization of the activation solution

Activator precursor (ml)	A	B	C	D
Colloidal Pd-Sn	25	50	62.5	82.5
HCl (37%)	62.5	50	50	83.5
DI-water	162.5	150	137.5	84

Acceleration of CNTs in a Mixture of Acids

The CNTs were then placed in a 50 g/L solution of a mixture of acids known commercially as the accelerator. The function of the accelerator is to remove excess tin hydroxide from the surface of the catalytic particles on CNTs allowing the palladium surface to be exposed. It was inferred that the acceleration does not remove tin from the core of the colloidal particles [7].

After acceleration, the CNTs were redispersed in water and filtered again to remove the traces of the previous solution. By following the previous steps, the surface of CNTs was rendered catalytic.

It was discovered that the best performance of the activation process comes at the composition C listed in Table 1 for each 0.1 gm of CNT's.

Electroless Cu-Co Plating of CNTs

A 1-liter solution of the Cu-Co electrolyte was prepared using the concentrations in Table 2.

All powders of the precursor were dissolved in DI water under sonication for 5 min. Formaldehyde is added to the solution after making sure of dissolving all the powders in the solution. Once the formaldehyde was added, Activated CNTs were put in the electroless bath under sonication for 2 min and the reaction started to take place on the catalytic surface of Pd-coated CNT's. Air bubbles started to come out from the solution once the CNTs were added. This happened as a sign of dissolving hydrogen from the surface of the palladium as well as the oxidation of formaldehyde that produces hydrogen. The solution was then put under magnetic stirring for 8 extra mins. Once the air bubbles stopped, it gave a good indication that copper has completely covered the entire catalyst surface. In this case the copper surface became an auto-catalyst and the solution turned dark brown indicating the coverage of CNTs by copper. After the stirring is completed, the copper coated CNT's started to accumulate in the bottom of the beaker due to their increased density. The solution was then filtered. The color of the filtered solution was monitored in the beginning of the filtration process and it was reported to be light pink indicating the consumption of all the copper ions in the solution before the filtration. Figure 2 shows the color of the obtained copper coated CNTs powder.

Once the electroless plating was optimized, the powder was ready for casting.

Table 2 Typical concentrations of Cu-Co electrolyte. [8]

Copper-Cobalt electrolyte	Concentrations
$Cu.SO_4.6H_2O$	6.99 g/L
Na_2CO_3	2 g/L
$CoCl_2.6H_2O$	1.09 g/L
$KNaC_2H_4O_6.4H_2O$ (Rochelle Salt)	22.57 g/L
NaOH	4.5 g/L
Formaldehyde 37%	2.59 ml/L

Fig. 2 Filtered copper coated CNT's of a brown color

Casting the Copper Coated CNT's Reinforced Aluminum Composite

CNTs begin to oxidize in air at 440–450 °C [9]. So, certain mechanical considerations were put to assure the protection of copper coated CNTs from oxidation. Using a flux to shield the surface of molten aluminum [10], the copper coated particles of different percentages were dispersed with a remelting flux in water via sonication. The mixture was then filtered. Commercial pure aluminum was melted in an induction furnace in a silicon carbide crucible. The temperature of the furnace was set at 850 °C. An infrared detector monitored the actual temperature of the molten liquid at 710 °C. Afterwards, the powder mixture of the copper coated CNT's and the flux was added to molten aluminum and stirred using a ceramic stirring rod. After removing the flux from the aluminum surface by a ceramic spoon, the molten aluminum was poured into a pre-heated steel mold at 300 °C and left to cool. The mould had a cylindrical cavity of 15 mm diameter and 16 cm length. The produced cylindrical rod was cut into slices using an isomet cutter. The sample was then grinded using sand papers from 400 to 2500 grit sizes. A silica suspension of 3 microns diameter was used in polishing the sample. All the samples were then tested on a Vickers hardness tester with 20 indentations per sample in different areas. The load was adjusted to 1 Kgf.

The copper coated CNTs were characterized using scanning electron microscopy (SEM) analysis using (LEO SUPRA 55VP FEG, Zeiss, equipped with Oxford EDS detector), transmission electron microscopy (TEM) using (JEM-2100 LaB6, JEOL, operating at 200 kV and equipped with Gatan SC200B CCD camera), energy dispersive X-ray (EDX) attached to the SEM, and X-ray diffraction (XRD) using (Cu Kα, Panalytical Xpert Pro diffractometer).

Results and Discussion

The activator composition C listed in Table 1 gave the best results when it comes to the copper coat conformity on top of CNTs. The addition of cobalt (II) chloride to the electroless copper solution helped in the process of autocatalytic reduction of copper ions in the electroless solution increasing the deposition rate tremendously. The use of colloidal Pd-Sn nanoparticles catalytic system that limited the catalyst optimization process to two factors (the concentration of the colloidal particles and volume of solution relative to the amount of CNTs) instead of four factors in the predecessor system. In addition, the new system provided a fixed average size of the colloidal nanoparticles for better coating conformity on top of CNTs. The optimized catalyst concentration and volume needed to cover the surface area of a fixed amount of CNTs helped in controlling the kick-start deposition rate of copper that is dependent on the catalyst concentration.

The SEM image in Fig. 3 shows CNTs after being coated with copper. The increase in the thickness of CNTs indicated the presence of copper on top of CNTs. The average diameter of the copper coated CNTs was found to be 85 nm compared to an average of 10–15 nm for as-received CNTs. The SEM image showed uniform coating that completely covers all the surfaces of CNTs.

As the TEM has a limited penetration depth, a sample of the copper coated CNTs was crushed using a mortar and pestle in order to remove some of the copper layers existing on top of the CNTs to differentiate dimensionally between the coated and uncoated part. Figure 4 shows the TEM and SEM images of the coated and uncoated parts of a MWCNT respectively. The TEM image shows that the copper coat is nanostructured.

Chemical analysis was done using energy dispersive X-ray (EDX) to determine the percentages of the elements present. Figure 5 shows the obtained spectrum.

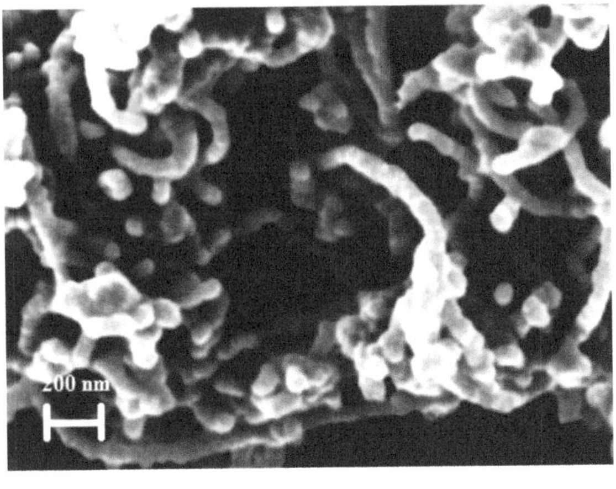

Fig. 3 SEM image of the copper coated CNT's

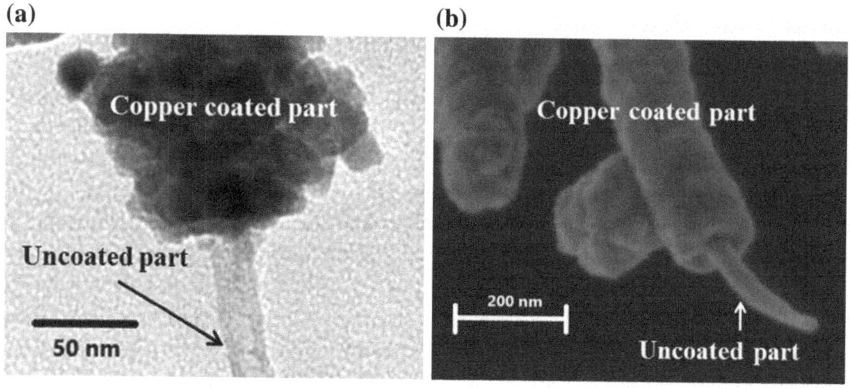

Fig. 4 **a** TEM Image showing the difference between copper coated and uncoated part of a MWCNT **b** An SEM image of the electroless copper coated and uncoated parts of a MWCNT

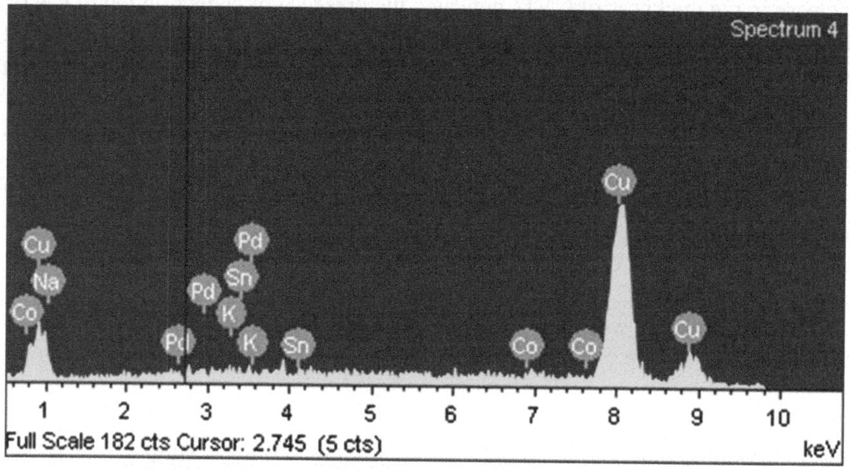

Fig. 5 EDX spectrum of copper coated CNTs

The analysis confirmed that 98.56% by weight of copper is obtained. Other elements such as Na, Pd, Sn and Co were found to exist in minimal percentages as listed in Table 3.

X-ray diffraction (XRD) confirmed the presence of copper (111), (200), and (220) crystallograpHic planes as shown in Fig. 6. Cu (II) oxide (111) plane appeared in the diffraction pattern indicating that a part of the sample was oxidized. It was clear that the (111) plane of CNTs was absent in the diffraction pattern. This can be attributed to the difference in the atomic weight between copper and carbon as well as the relatively small amounts of carbon present. Other researchers who have prepared copper CNTs also reported the absence of CNT peak expected at 26°

Table 3 EDX elemental analysis of copper coated CNT's

Element	Weight (%)	Atomic (%)
Cu	98.56	98.32
Na	0.2	0.55
Co	0.66	0.71
K	0.08	0.13
Pd	0.43	0.26
Sn	0.07	0.04
Total	100%	

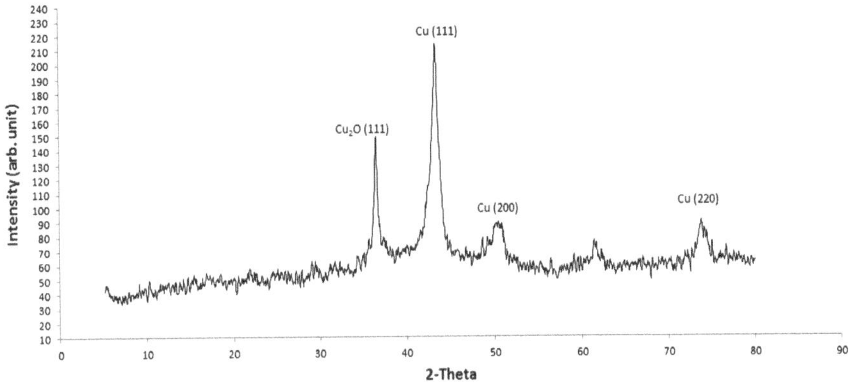

Fig. 6 XRD pattern of Copper coated CNTs

in the XRD diffraction pattern [11]. However, the pattern obtained in the present study showed much stronger copper (111) peak that was higher than the Cu_2O peak unlike what other researchers found [11]. This means that the present sample was less likely to oxidize.

As mentioned earlier, the copper coated CNTs were added in different percentage to pure aluminum using the casting technique. The purpose from the addition is to improve the wettability and dispersion between aluminum and CNTs that was reported very poor in previous reviews [12]. The hardness of the cast samples was tested using Vickers hardness tester. About 20 different indentations were conducted in different areas of each sample. The small variability of Vickers hardness between different indentations was an indicative of the homogeneous distribution of Cu-coated CNTs in the aluminum matrix. The results of the Vickers hardness number and standard error (SE) for the different copper CNTs percentages are presented in Table 4.

It was found that adding 0.5, 1 and 2% of the coated CNTs to pure aluminum resulted in 12.2, 25.6, and 52.1% increase in Vickers hardness which is significant especially that, for example, the 2% copper coated CNTs has less than 0.1% of MWCNTs. This confirms the potential of the process used in producing good

Table 4 Vickers hardness number of casted aluminum having different percentages of copper coated CNTs

Cu-CNTs % by weight	Vickers hardness value HV_1 at 1000 g load
Pure Al	36.6 (SE = 1.5)
Al + 0.5% Cu-CNTs	41.1(SE = 1.64)
Al + 1% Cu-CNTs	49.2 (SE = 1.67)
Al + 2% Cu-CNTs	76.42 (SE = 1.7)

quality cast composites of Al-CNT. Efforts are underway to optimize the casting process and to fully investigate the mechanical behavior of the composites.

Conclusions

One of the challenges facing electroless plating of copper on CNT's is the rapid growth of copper in uncontrolled manner. The results of the current study showed the success of using colloidal Pd-Sn besides the copper cobalt electrolyte in reaching an optimized electroless plating process of an end point of the reaction controlled in a short interval of time due to the increase of the kinetics of the reaction. This mechanism allowed the prepared copper coated CNT's to be filtered without the fear of copper being coated on the accumulated copper CNT's on the filter membrane. The SEM and TEM images confirmed the uniformity of the coat without the presence of copper crystals and aggregates in the produced structure. The use of CNTs without acidic functionalization added a plus as it shortened the processing time without adding extra cost and without affecting the dispersion of CNT's significantly. Furthermore, the copper coated CNTs resulted in composites of uniform properties when mixed with aluminum during casting as confirmed by the hardness test results. The addition of the electroplated Cu-CNTs resulted in increasing the Vickers hardness of pure Al by up to 52.1% even when the CNT's percent was less than 0.1% thus confirming the potential of this composite for applications requiring improved mechanical properties combined with the shape complexities facilitated by using a casting technique. This can have important commercial applications especially that the activation system is commercialized in other industries such as the plating on plastics industry, which adds to the applicability of the process to be upscaled in the metal industry. The overall strategy of the study followed a tremendous reduction in the production cost for encouraging different industries to adapt such a technique.

Acknowledgements The authors wish to thank EgyptAlum Company in Nagaa Hammady, Egypt for donating the commercial pure aluminum samples for the purpose of research and development. M. Elsharkawi is grateful for the financial support by the Office of the Dean of Graduate Studies and Research at the American University in Cairo and by Al-Alfi foundation.

References

1. Esawi A, Morsi K, Sayed A et al (2010) Effect of carbon nanotube (CNT) content on the mechanical properties of CNT-reinforced aluminium composites. Composites Sci Tech 70:2237–2241. https://doi.org/10.1016/j.compscitech.2010.05.004
2. Esawi A, Morsi K, Sayed A et al (2009) Fabrication and properties of dispersed carbon nanotube–aluminum composites. Mater Sci Eng, A 508:167–173. https://doi.org/10.1016/j.msea.2009.01.002
3. So KP, Lee IH, Duong DL et al (2011) Improving the wettability of aluminum on carbon nanotubes. Acta Mater 59:3313–3320. https://doi.org/10.1016/j.actamat.2011.01.061
4. Mallory GO, Hajdu JB (2009) Electroless plating: fundamentals and applications. Am Electroplaters Surf Finish, New York
5. Feng Y, Yuan H (2004) Electroless plating of carbon nanotubes with silver. J Mat Sci 39:3241–3243. https://doi.org/10.1023/b:jmsc.0000025869.05546.94
6. Mishra KG (1996) Kinetics and mechanism of electroless deposition of copper. J Electrochem Soc 143:510. https://doi.org/10.1149/1.1836473
7. Cohen R, Meek R (1976) The chemistry of palladium—tin colloid sensitizing processes. J Colloid Interface Sci 55:156–162. https://doi.org/10.1016/0021-9797(76)90021-7
8. Shacham-Diamand Y, Sverdlov Y, Friedberg S, Yaverboim A. (2017) Electroless plating and printing technologies, in nanomaterials for 2D and 3D printing. In: S. Magdassi and A. Kamyshny (eds) Wiley-VCH Verlag GmbH & Co. KGaA, Weinheim, Germany. https://doi.org/10.1002/9783527685790.ch3
9. Osswald S, Havel M, Gogotsi Y (2007) Monitoring oxidation of multiwalled carbon nanotubes by Raman spectroscopy. J Raman Spectrosc 38:728–736. https://doi.org/10.1002/jrs.1686
10. Mansour M, Shahid M (2016) Carbon nanotube-reinforced aluminum composite produced by induction melting. JART 12(4):215–224. https://doi.org/10.1016/j.jart.2016.05.002
11. Singhal SK, Lal M, Sharma I, Mathur RB (2012) Fabrication of copper matrix composites reinforced with carbon nanotubes using a combination of molecular-level-mixing and high energy ball milling. J Compos Mater 47:613–621. https://doi.org/10.1177/0021998312443397
12. Agarwal A, Bakshi SR, Lahiri D (2011) Processing techniques. Carbon nanotubes: reinforced metal matrix composites. CRC Press-Taylor & Francis, Boca Raton, Florida, pp 30–33

Part II
Synthesis and Developments of Emerging Composites

Study on Hot Deformation Behavior and Processing Map of 20 Volume Percent $Al_{18}B_4O_{33w}$/2024 Composites

Wenchen Xu, Xiangqian Zeng, Xueze Jin and Debin Shan

Abstract The hot deformation behavior and processing map of 20vol% $Al_{18}B_4O_{33w}$/2024 composites were investigated by isothermal compression test within the temperature range of 380–500 °C and the strain rate range of 0.001–1 s^{-1}. The results show that with the increase of temperature and decrease of strain rate, the flow stress decreased gradually during hot deformation. Due to dispersed strengthening caused by $Al_{18}B_4O_{33}$ whiskers, the activation energy of the composite was much higher than the self-diffusion energy of pure Al. With the increase of deformation temperature, the deformation softening mechanism of the Al matrix composites was controlled both by dynamic recovery and dynamic recrystallization. The dynamic material model (DMM) was reliable in predicting the safe deformation regions. The Prasad's instability criterion was appropriate in predicting the flow instability regions. The optimum deformation condition was 460–480 °C and 0.01–0.1 s^{-1}, which was validated by near isothermal two-directional forging experiment. This study provided an effective guidance on hot working of $Al_{18}B_4O_{33w}$ whisker reinforced Al matrix composites.

Keywords Aluminum matrix composite · $AL_{18}B_4O_{33}$ whisker
Hot compression · Mechanical properties

Introduction

Aluminum matrix composites (AMCs) are increasingly used in aviation, aerospace and automobile industries due to their low density, high specific strength, high specific modulus, good wear resistance and thermal conductivity [1–3]. In the early

W. Xu (✉) · X. Zeng · X. Jin · D. Shan (✉)
National Key Laboratory for Precision Hot Processing of Metals, School of Materials Science and Engineering, Harbin Institute of Technology, Harbin 150001, People's Republic of China
e-mail: xuwc_76@hit.edu.cn

D. Shan
e-mail: shandebin@hit.edu.cn

T. S. Srivatsan et al. (eds.), *Metal-Matrix Composites Innovations, Advances and Applications*, The Minerals, Metals & Materials Series,
https://doi.org/10.1007/978-3-319-72853-7_4

stage, SiC whiskers and Si_3N_4 whiskers were selected as important reinforcements to strengthen aluminum matrix composites. However, the two kinds of whiskers were not widely used owing to their high production cost. In 1976, the $Al_{18}B_4O_{33}$ whiskers were developed in Japan, which possessed high strength and Young's modulus [4]. Compared to the SiC whiskers and Si_3N_4 whiskers, the $Al_{18}B_4O_{33}$ whiskers were much cheaper and easier to prepare, showing great application potential in industry application.

Currently, the whisker reinforced aluminum matrix composites (AMCs) can be fabricated by powder metallurgy, squeeze casting, stirring casting, pressure infiltration, etc. However, the existence of amounts of cavities and the clustering effect of whiskers may decrease the mechanical properties of aluminum matrix composites, which limits their wide application in various fields [5]. Although subsequent hot working processes, such as hot rolling [6, 7], hot extrusion [8, 9] and isothermal forging [10], can improve the microstructure and mechanical properties, the introduction of low-ductility whiskers significantly reduced the workability of the AMCs. During hot deformation, some defects such as surface cracking and whisker fracture, are extremely easy to occur, which makes the AMCs difficult to be plastically deformed. Nevertheless, the hot processing parameters, such as temperature, strain rate and deformation degree, have important influence on the deformation behavior of AMCs. Especially, deformation softening, such as dynamic recrystallization and recovery, may improve the hot workability of AMCs [11]. Therefore, it is necessary to conduct symmetric study on the hot deformation behavior and optimize the hot working window, which is helpful for the hot processing and application of whisker reinforced AMCS.

In this study, the hot deformation behavior of $20vol\%Al_{18}B_4O_{33w}/2024Al$ composites under different temperatures and strain rates was investigated by isothermal compression tests and the dynamic softening mechanism was analyzed. In order to optimize the hot deformation window, the processing maps were established based on Prasad's instability criterion and the microstructure evolution was discussed as well. The reliability of hot working window was verified by near isothermal two-directional forging experiment.

Experimental

The 2024Al matrix composites reinforced with $20vol\%Al_{18}B_4O_{33w}$ whisker were fabricated by squeeze casting technique, wherein the whisker diameter was about 0.5–1 μm, and the length was about 10–20 μm. The cylindrical specimens with 8 mm in diameter and 12 mm in height for thermal compression test were spark machined from the casting ingot. The isothermal compression experiment was carried out on a Gleeb-1500D thermal simulator within the strain rate range of 0.001–1 s^{-1} at the temperatures of 380, 400, 420, 440, 460, 480 and 500 °C, respectively. In order to diminish the influence of friction, the graphite flake was placed on both the top and bottom sides of the cylindrical specimen, used as lubricant. In the hot compression experiment, the specimens were heated to the test temperature at the rate of 10 °C/s, kept for 3 min, and then compressed with the height reduction of 50%, followed by quenching into the water.

After hot compression, the specimens were sectioned for microstructural observation using an Olympus-PMG3 optical microscope (OM). The SEM observation was performed using a Quanta 200FEG scanning electron microscope equipped with energy-dispersive spectroscopy (EDS). The TEM analysis were conducted on a Talos F200x transmission electron microscope operated at 200 kV, and the specimens for TEM observation were made by ion milling system.

Results and Discussion

Initial Microstructure

Figure 1a, b show the micrographs of as-cast 20vol%$Al_{18}B_4O_{33w}$/2024Al composites. It is obvious that the whiskers are distributed uniformly and randomly in the Al matrix composites. Besides, the white particles represents the secondary phases ($CuAl_2$), as indicated by the blue arrows. In addition, some cavities with relatively larger size existed in the as-cast composite, especially at the interface of $Al_{18}B_4O_{33}$w and the Al matrix or the secondary phases, as indicated by the red arrows, as can be seen in Fig. 1b.

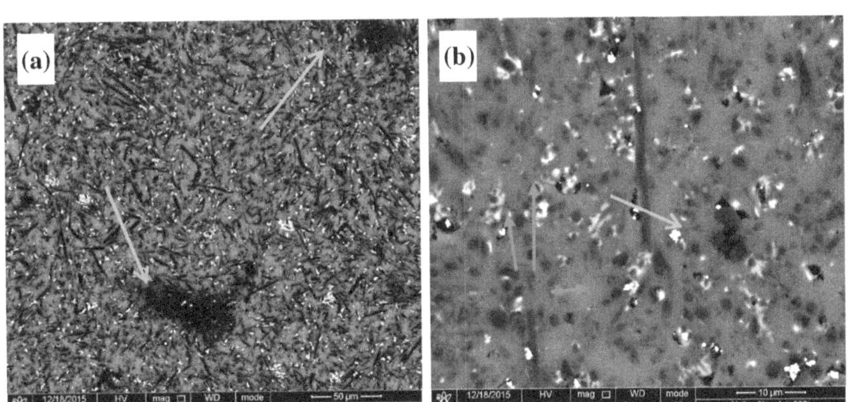

Fig. 1 SEM Micrographs of as-cast $Al_{18}B_4O_{33}$w/2024Al

Flow Stress-Strain Behavior and Macroscopic Morphologies

Flow Stress Curves

Although the graphite sheet was used as lubricant during hot compression, the thermal friction between the specimen and anvil could not be ignored, thus it is necessary to implement the friction correction to decrease the influence of uneven deformation on the compression pressure. The exponential friction correction method was adopted in this study, which was deduced by Ebrahimi et al. [12–14]. The corrected flow strain-stress curves of 20vol%$Al_{18}B_4O_{33w}$/2024Al composite under the different strain rates and temperatures are presented in Fig. 2. At the initial stage, the flow stress grew rapidly to a peak value because of work hardening. Beyond the peak stress, the stress–strain curves exhibited a gradual decline with increasing strain. And then the flow stress entered into a relatively steady stage, especially at high temperature and strain rate.

With the increase of deformation temperature, dynamic softening enhanced and dominated the flow stress in the steady deformation stage. For the Al matrix composites, the predominated softening mechanism during hot compression was possibly whisker rotation [15, 16], dynamic recovery, and dynamic recrystallization [17].

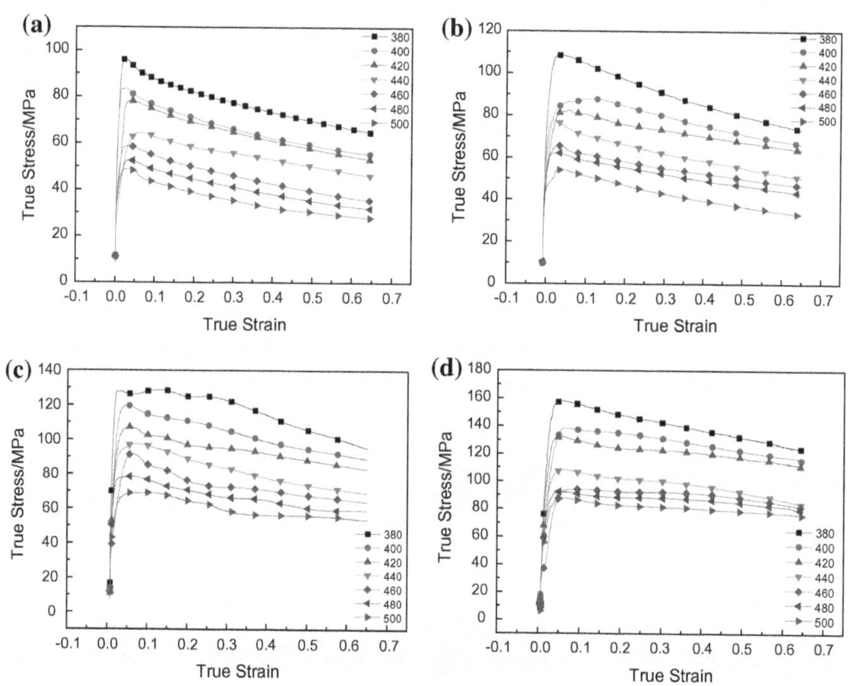

Fig. 2 Flow strain-stress curves of 20vol%$Al_{18}B_4O_{33w}$/2024Al composite in different strain rates **a** 0.001 s^{-1}, **b** 0.01 s^{-1}, **c** 0.1 s^{-1}, **d** 1 s^{-1}

Kinetic Analysis

As two important kinetic parameters, the strain-rate sensitivity coefficient m and apparent activation energy Q are widely used to describe the intrinsic deformation behavior of metals and MMCs [18, 19]. Generally, the strain rate, flow stress and deformation temperature of metals and MMCs could be expressed by hyperbolic sine function [20]:

$$\dot{\varepsilon} = A[\sinh(\alpha\sigma)]^n \exp(-\frac{Q}{RT}) \tag{1}$$

where A and α are the materials constants independent of temperature, n is the stress exponent, $\dot{\varepsilon}$ is strain rate, T is deformation temperature, R is the ideal gas constant and Q is the thermal activation energy. Based on the equations above, the stress exponent n and activation energy Q can be defined as:

$$\frac{1}{n} = \frac{\partial \ln[\sinh(\alpha\sigma)]}{\partial \ln \dot{\varepsilon}} |T \tag{2}$$

$$Q = R\{\frac{\partial \ln \dot{\varepsilon}}{\partial \ln[\sinh(\alpha\sigma)]}\}T\{\frac{\partial \ln[\sinh(\alpha\sigma)]}{\partial(1/T)}\}\dot{\varepsilon} \tag{3}$$

The linear relationship of strain rate and flow stress was shown in the Fig. 3a, b. According to Eqs. (1) and (2), the α was calculated as 11.627×10^{-3}. Figure 3c, d present the linear relationship of $\ln[\sinh(\alpha\sigma)] - \ln \dot{\varepsilon}$ and $\ln[\sinh(\alpha\sigma)] - 1/T$, respectively. According to Eq. (3), the activation energy Q was calculated as 286.9 kJ/mol. According to the linear relationship of $\ln[\sinh(\alpha\sigma)] - \ln Z$ shown in the Fig. 3e, the slope and intercept of the line denoted the stress exponent n and the constant A, which were estimated as 9.798 and 5.326×10^{18}, respectively. The linear correlation coefficient r was 0.99275, indicated it was reliable to describe the hot deformation behavior of the Al matrix composites by hyperbolic sine law. Thus, the constitutive equation of 20vol%$Al_{18}B_4O_{33}$w/2024Al composite with substitutions of A, α, n and Q *into* Eq. (1) could be obtained as follows:

$$\dot{\varepsilon} = 5.326 \times 10^{18}[\sinh(8.4956 \times 10^{-3}\sigma)]^{9.978} \exp(-\frac{286900}{8.314T}) \tag{4}$$

The activation energy of composites was much higher than the self-diffusion activation energy of Al which is only 142 kJ/mol [21]. Since the stress exponent n $\geq$ 8, the main strengthening mechanism was possibly dispersion strengthening. Hence, the existence of reinforced whisker and hard phase $CuAl_2$ could impede dislocation motion, leading to the strength increase of the Al matrix composites.

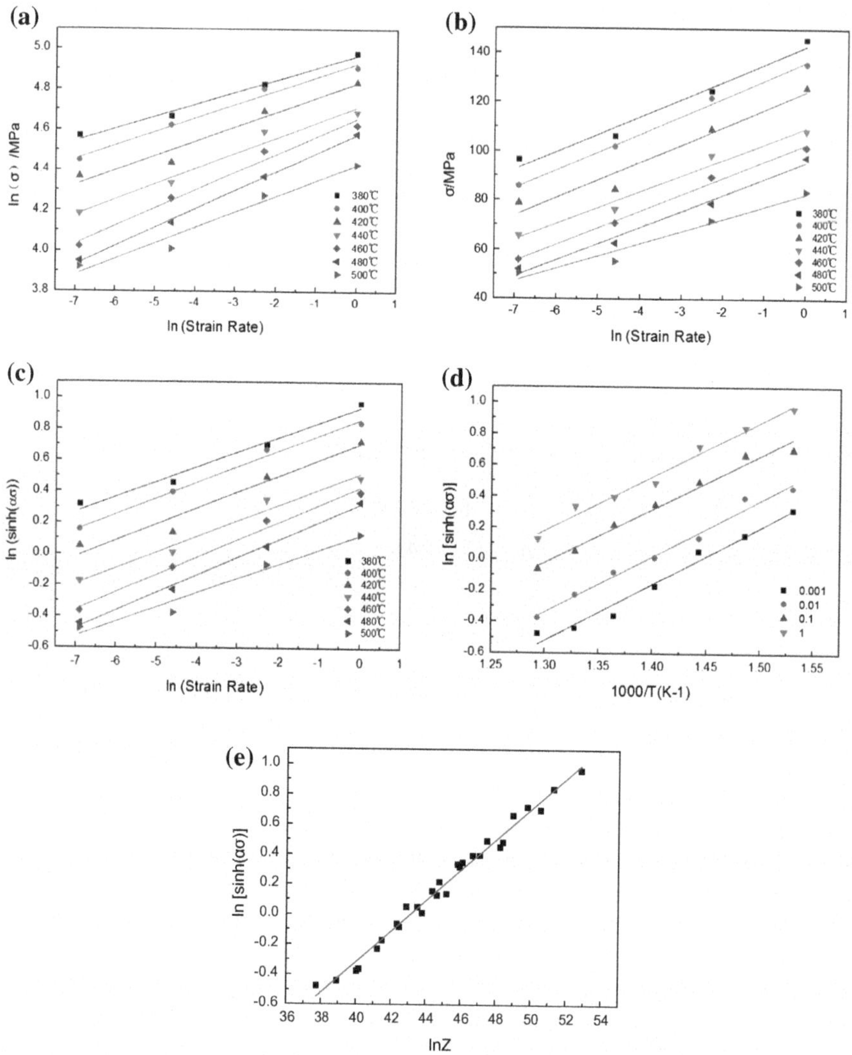

Fig. 3 Relationships of the 20vol%$Al_{18}B_4O_{33w}$/2024Al composites between the peak stress: **a** $\ln \sigma - \ln \dot{\varepsilon}$, **b** $\sigma - \ln \dot{\varepsilon}$, **c** $\ln[\sinh(\alpha\sigma)] - \ln \dot{\varepsilon}$, **d** $\ln[\sinh(\alpha\sigma)] - 1/T$, **e** $\ln[\sinh(\alpha\sigma)] - \ln Z$

Macroscopic Morphology

Table 1 shows the macroscopic morphology of compressed specimens with the true strain of 0.69 at different strain rates and temperatures. Obviously, cracks occurred when the samples were compressed at 380 °C and higher strain rate ($\geq 0.1\ s^{-1}$) because deformation softening induced through dynamic recovery or dynamic recrystallization was insufficient, and the dislocation pile-up became more severe

Table 1 Macroscopic morphologies of hot compressed specimens at different temperatures and strain rates with true strain of 0.69

Strain rate/s^{-1}	Temperature/°C						
	380	400	420	440	460	480	500
0.001							
0.01							
0.1							
1							

with the progress of compression deformation at lower temperature. At higher deformation temperature, the cracking tended to reduce due to more sufficient dynamic recovery or dynamic recrystallization. As shown in Table 1, it is obvious that under higher temperature, the macroscopic morphology of specimens at higher strain rates ($\geq 0.1\ s^{-1}$) was better than that of specimens at low strain rates, which indicates that in this region the 20vol%$Al_{18}B_4O_{33}w$/2024Al composites possessed better workability. However, the cracks appeared at the deformation temperature of 500 °C and strain rate of 1 s^{-1}. The possible cause is that, the secondary phase $CuAl_2$ in the 2024 matrix was actually an eutectic phase with the melting temperature of about 505 °C [22], so the deformation temperature of 500 °C was high enough to induce cracking in the vicinity of $CuAl_2$ phases during hot compression of the 20vol%$Al_{18}B_4O_{33}w$/2024Al composites. Moreover, higher strain rate led to high deformation heat accumulation [23], which increased the actual deformation temperature higher than 500 °C in the AMCs. Therefore, the main region of hot compressed specimens without cracks was mainly located at high temperatures and medium to high strain rates.

Processing Maps and Analysis

Basic Theory of Processing Map

It has been authenticated that processing map is an effective way to analyze the hot deformation behavior and obtain the hot working parameters. The processing map is based on the dynamic material model proposed by Prasad et al. [24]. In this model, the material is considered as a nonlinear power dissipation unit. During hot deformation, the input energy *P* in every moment is divided into two parts, i.e. the power dissipation *G* which was the consumed energy and the power dissipation correlation *J* which was caused by microstructure evolution like dynamic recovery, dynamic recrystallization and internal cracks. According to the dynamic material model, the energy absorbed by material *P* can be presented through flow stress σ and strain rate $\dot{\varepsilon}$ as:

$$P = \sigma \cdot \dot{\varepsilon} = G + J = \int_0^{\dot{\varepsilon}} \sigma d\dot{\varepsilon} + \int_0^{\sigma} \dot{\varepsilon} d\sigma \tag{5}$$

Under certain temperature and deformation, the relationship of flow stress and strain rate can be considered as:

$$\sigma = K\dot{\varepsilon}^m \tag{6}$$

where K is a constant, m is strain rate sensitivity index that is the ratio of two parts of energy:

$$m = \frac{dJ}{dG} = \frac{\partial \ln \sigma}{\partial \ln \dot{\varepsilon}} \tag{7}$$

The power dissipation correlation J can be defined as:

$$J = \int_0^\sigma \dot{\varepsilon} d\sigma = \frac{m}{m+1} \sigma \dot{\varepsilon} \tag{8}$$

When the value of m is 1, it means the material is under ideal linear dissipative state, thus the maximum value of J is $J_{\max} = \frac{1}{2} \sigma \dot{\varepsilon}$.

To describe the dissipative state of material under different temperatures and strain rates, the power dissipation coefficient η is proposed as:

$$\eta = \frac{J}{J_{\max}} = \frac{2m}{m+1}. \tag{9}$$

Power Dissipation Maps

The energy dissipation maps at the true strain of 0.5 and 0.69 were shown in Fig. 4. The number of contour line represented the value of the power dissipation coefficient η. As seen from Fig. 4a, b, when the strain was 0.5, the power dissipation coefficient of the composites increased from the lower left corner to the upper right corner, and there are two peak regions: the first region appeared within the deformation temperatures of 470–490 °C and the strain rates of 0.001–0.01 s^{-1}; the second region occurred within the deformation temperatures of 480–500 °C and the strain rates of 0.032–1 s^{-1}.

It could be observed from Fig. 4c, d that there are two power dissipation efficiency peak regions at the strain of 0.69. The first region appeared at the deformation temperature of 460–485 °C and the strain rate of 0.001–0.032 s^{-1}, and the second region occurred at the deformation temperature of 490–500 °C and the strain rate of 0.05–0.32 s^{-1}. The peak dissipation efficiency coefficients of these two regions are 0.25, indicating that microstructural evolution like dynamic recovery and dynamic recrystallization may appear in these two regions. In these two power dissipation maps, all the peak power dissipation values appeared at the deformation temperature of 465–485 °C and lower strain rate. It is generally believed that the power dissipation coefficient between 0.25 and 0.5 indicated the occurrence of dynamic recrystallization, so the main softening mechanism of the composites during the hot working process may be dynamic recovery and dynamic recrystallization at the higher temperature and lower strain rate. Therefore, the composites show better workability in this area (Fig. 4).

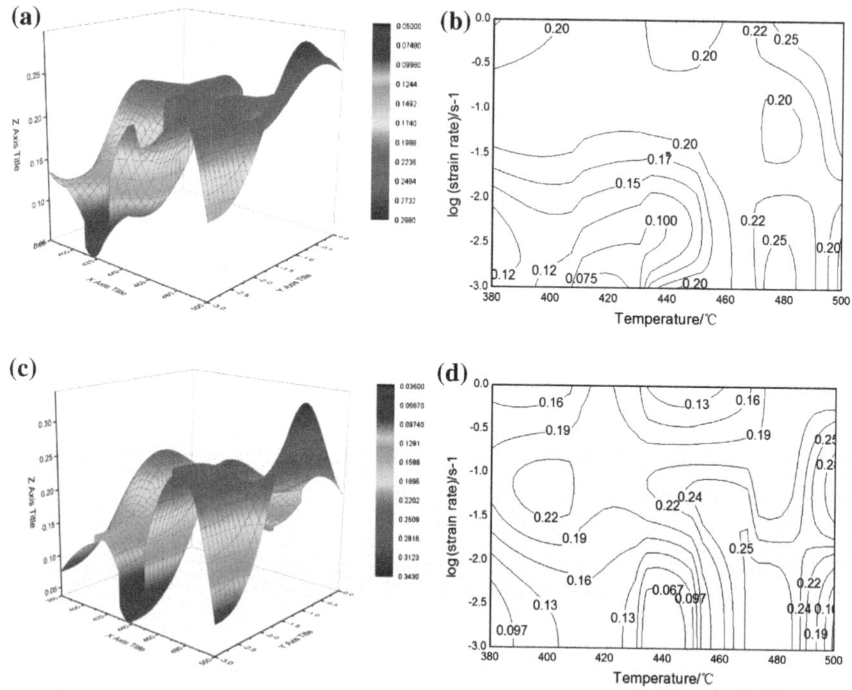

Fig. 4 The energy dissipation diagram at the true strain of 0.5 and 0.69 of 20vol%$Al_{18}B_4O_{33}$w/2024Al composites. **a** Energy dissipation diagram at the true strain of 0.5. **b** Contour line projection of energy dissipation diagram at the true strain of 0.5. **c** Energy dissipation diagram at the true strain of 0.69. **d** Contour line projection of energy dissipation diagram at the true strain of 0.69

Hot Processing Maps

Figure 5 shows the hot processing maps at the true strain of 0.69 based on DMM, wherein the red shadow areas represent flow instability zones. It could be found that plastic deformation instability mainly occurred in high strain rate region of 0.1–1 s^{-1}. Because at high strain rates, the dislocation density increased during hot deformation, deformation softening could not proceed completely. Severe plastic deformation at higher strain rates may cause crack initiation. Therefore, this area should be avoided when the composites are forged.

In the safe processing area of the composites, the greater the power dissipation efficiency coefficient η was, the lower the energy state of composite was and the better the plastic deformability was. According to Fig. 5, the optimal processing area of 20vol%$Al_{18}B_4O_{33}$w/2024Al composites should be located in the region with deformation temperature of 460–485 °C and strain rate of 0.001–0.032 s^{-1}.

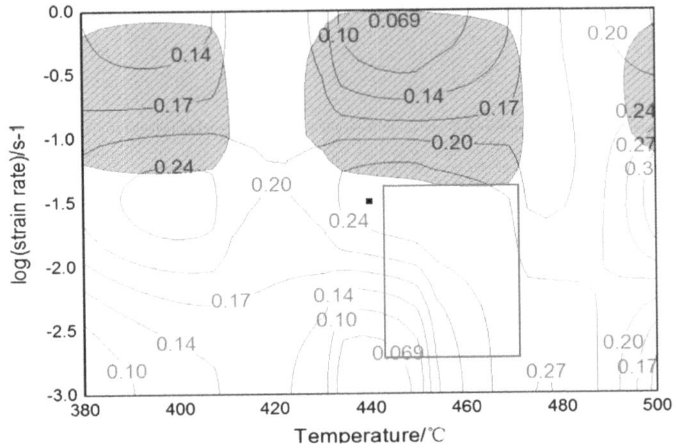

Fig. 5 The processing maps based on Prasad's instability criterion at the true strain of 0.69 of 20vol%SiC_w/6061Al composites

Deformation Microstructure in Different Regions

Figure 6 shows the microstructures of 20vol%$Al_{18}B_4O_{33}$w/2024Al composites compressed at different temperatures and strain rates. Obviously, compared to the as-cast microstructure of composites, the cavities with relative larger size disappeared after hot compression. When the samples were compressed at 400 °C/0.1 s^{-1}, some small cavities occurred at the interface of $Al_{18}B_4O_{33}$ whiskers and 2024Al alloy matrix, and there were some cracks visible in local regions, as shown in Fig. 6a). As the temperature increased to 460 and 480 °C, few cavities occurred in the microstructure and the rotation of whickers became more obviously, which would be conducive to plastic deformation. During hot compression, the stress was exerted in the matrix and then transferred to the $Al_{18}B_4O_{33}$ whiskers and the interface between the Al matrix and $Al_{18}B_4O_{33}$ whiskers. Therefore, the stress concentration would occur easily at the interface of Al matrix and $Al_{18}B_4O_{33}$ whiskers. When the value of stress concentration exceeded the interface bonding strength, cracks developed along the interface. With the increase of deformation temperature, the rotation of matrix grains and whiskers became more flexible. Besides, at high deformation temperature, dynamic softening including dynamic recovery and dynamic recrystallization proceeded more sufficiently. Therefore the stress concentration at the interface became weaker, which decreased the probability of cracking at the interface.

The TEM images of samples deformed at high temperature and moderate strain rate (480 °C/0.01 s^{-1}) were shown Fig. 7. Figure 7a depicted dislocations piled-up near the grain boundaries and the dynamic recovery (DRV). As the stacking fault energy of 2024Al alloy was relatively high, the predominated softening mechanism should be dynamic recovery (DRV). However, dynamic recrystallization also

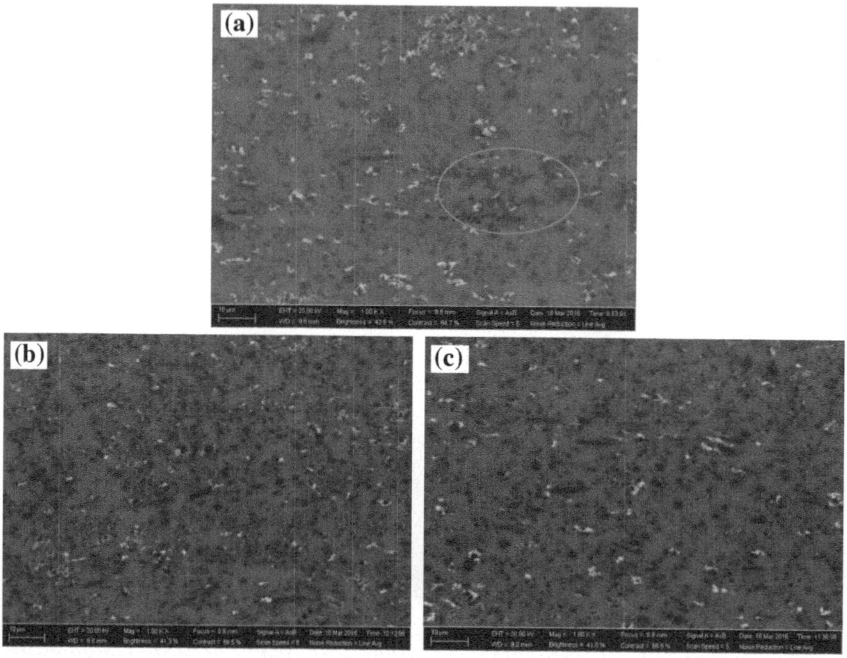

Fig. 6 Microstructures of 20vol%$Al_{18}B_4O_{33w}$/2024Al deformed to a true strain of 0.69 at different temperature and strain rates. **a** 400 °C/0.1 s^{-1}, **b** 460 °C/0.01 s^{-1}, **c** 480 °C/0.01 s^{-1}

occurred during hot compression, as shown in Fig. 7c. Because the dislocation piled up near the grain boundaries and whiskers, the dislocation density increased rapidly to promote the nucleation of DRX. Due to the occurrence of DRX, the plastic workability of Al composites was improved and the uneven deformation became weaker. Therefore, both dynamic recovery and dynamic recrystallization were the deformation softening mechanism of 20vol%$Al_{18}B_4O_{33w}$/2024Al composites during hot compression.

Application and Verification of Processing Maps

As the optimal deformation parameters including temperature and strain rate were obtained from the processing map, the near isothermal two-directional forging experiment was carried out at a starting temperature of 470 °C and the load speed of 1 mm/s (corresponding to the strain rate of 0.032 s^{-1}). The initial cylindrical billet was 47 mm in diameter and 50 mm in height. Figure 8a shows the macroscopic morphology of the AMCs subject to two-directional forging. Figure 8b, c show the microstructure of the forged composites in two directions, i.e. in the plane parallel and vertical to initial compression axis, respectively. Obviously, the composites

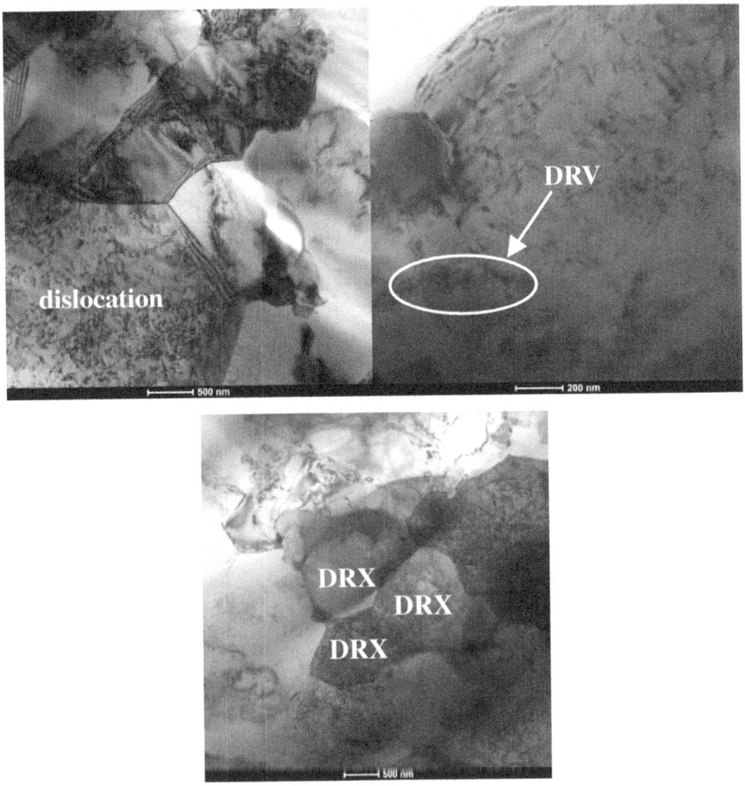

Fig. 7 TEM micrograph of $Al_{18}B_4O_{33}$w/2024Al composites compressed at 0.01 s^{-1} and 480 °C showing: **a** dislocation, **b** dynamic recovery, and **c** dynamic recrystallization

forging exhibited perfect surface without evident deformation defects. After multi-pass forging in two directions, it could be observed that the distribution of whickers became more uniform and few cavities were found in the planes parallel and vertical to compression axis. Figure 9 shows the comparison of the mechanical properties of the AMCs before and after two-directional forging. It is obvious that the strength and elongation of as-forged composites were much higher than that of as-cast composites. After two-directional forging, the tensile strength of the composites increased from 240 MPa to about 330 MPa, and the elongation increased from 0.8 to 3.0%. The tensile strength of the forged composites was enhanced by 90 MPa, and the elongation was increased by nearly 275%, which indicated significant improvement of mechanical properties of the AMCs and verified the reliability of hot working zone predicted by the processing map in this study.

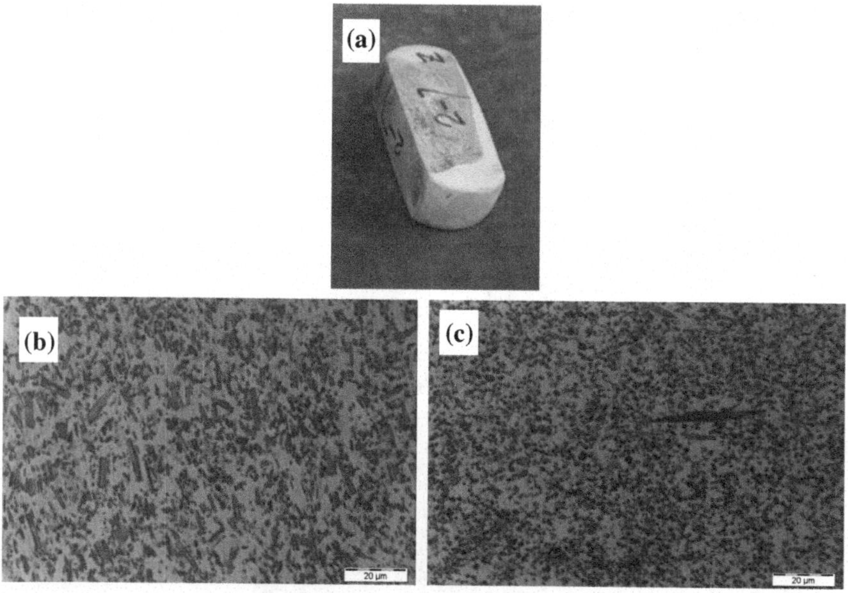

Fig. 8 The image and microstructure of 20vol%$Al_{18}B_4O_{33}$w/2024Al composite forging after two-directional forging. **a** Photo of composites forgings **b** OM microstructure of as-forged sample in the plane vertical to initial compression axis **c** OM microstructure of as-forged sample in the plane parallel to initial compression axis

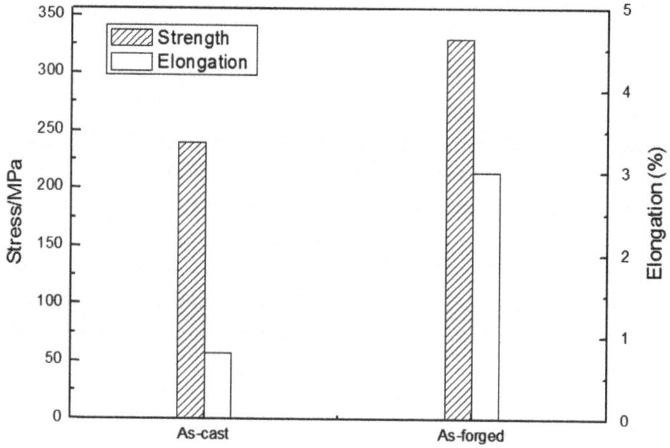

Fig. 9 The tensile strength and elongation of as-cast and forged composites

Conclusions

Through investigating the hot compressive behavior of 20vol%$Al_{18}B_4O_{33}$w/2024Al composites in the temperature range of 380–500 °C and strain rate range of 0.001–1 s^{-1}, the main conclusions could be obtained as follows:

(1) During the hot compression of the AMCs, the flow stress decreased with the increase of temperature and decrease of strain rate. Under the condition of low temperature and high strain rate, the AMCs were more likely to be damaged, while under high temperature and medium strain rate, dynamic softening and the rotation of the whiskers were prone to initiate to coordinate the plastic deformation in the AMCs.
(2) The apparent activation energy Q of the AMCs was determined as 286.90 kJ/mol, which was far higher than the self-diffusion energy of Al (142 kJ/mol) because of the dispersed strengthening of $Al_{18}B_4O_{33}$ whiskers. The constitutive equation for hot deformation was established as:

$$\dot{\varepsilon} = 5.326 \times 10^{18}[\sinh(8.4956 \times 10^{-3}\sigma)]^{9.978}\exp(-\frac{286900}{8.314T}).$$

(3) The safe deformation region of the 20vol%$Al_{18}B_4O_{33}$w/2024Al composites was 460–490 °C and 0.001–0.032 s^{-1}. In the safe deformation region, the occurrence of dynamic recovery and dynamic recrystallization reduced the deformation resistance and the stress concentration at the interface between $Al_{18}B_4O_{33}$ whiskers, which effectively avoided the nucleation of cracks and improved the workability of the AMCs during hot deformation.
(4) The two-directional forging experiment verified the reliability of the safe deformation region predicted by the processing map established in this study. After two-directional forging, the AMCS shows significant improvement of the mechanical properties, the tensile strength and elongation increasing by 37.5 and 275%, respectively.

References

1. Fan J, Sang J, Shi L (2001) Fabrication, application and development of particle reinforced aluminum matrix composites. Mater Rev 15(10):55–58
2. Hao B, Duan X, Cui H et al (2005) Present status and expectation of metal matrix composites. Mater Rev 19(7):64–68
3. Prasada SV, Asthana R (2004) Aluminum metal-matrix composites for automotive applications: tribological considerations. Tribol Lett 17(3):445–453
4. Zhang X (2013) The discussion of $Al_{18}B_4O_{33w}$/Al composites. Equip Manuf Technol 11: 89–90
5. Zhang P, Li F, Li H (2009) Optimization of general constitutive equation for hot deformation of SiC particle reinforced Al matrix composites. J Aeronaut Mater 29(1):51–56

6. Fei WD, Li WZ, Yao CK (2002) Hot rolling behaviors of whisker reinforced aluminum composites. J Mater Sci 37(1):211
7. Lee JC, Subramanian KN (1996) Tensile properties of hot-rolled (Al_2O_3)p/Al composites. Mater Sci Eng, A 196(1–2):71
8. Standford-Beale CA, Clyne TW (1989) Extrusion and high-temperature deformation of fibre reinforced aluminum. Compos Sci Technol 35(2):121–157
9. Ganesh VV, Chawla N (2005) Effect of particle orientation anisotropy on the tensile behavior of metal matrix composites: experiments and microstructure-based simulation. Mater Sci Eng, A 391(1–2):342–353
10. Chu WG, Hu J, Fei WD, Yao CK (1999) Microstructure evolution of isothermally-forged SiCw/Al composite. J Mater Sci 34(3):565
11. Fei WD, Li C, Jiang XD (1993) Rotation of whiskers and its effect on high temperature compressive floe stress of SiCw/Al-Ni composite. Acta Metall Sin 20:283
12. Ebrahimi R, Najafizadeh A (2004) A new method for evaluation of friction in bulk metal forming. J Mater Process Technol 152:136–143
13. Zhang C, Shen LW, Shen WF, Liu CR, Xia YN, Li RQ (2016) Study on constitutive modeling and processing maps for hot deformation of medium carbon Cr-Ni-Mo alloyed steel. Mater Des 90:804–814
14. Dong YY, Zhang CS, Zhao GQ, Guan YJ, Gao AJ, Sun WC (2016) Constitutive equation and processing maps of an Al–Mg–Si aluminum alloy: determination and application in simulating extrusion process of complex profiles. Mater Des 92:983–997
15. Li ZJ, Fei WD, Wang LD (2006) Hot compressive deformation behavior of aluminum matrix composite reinforced by $Al_{18}B_4O_{33}$ whisker coated with Bi_2O_3. Mater Sci Eng, A 432: 275–280
16. Kalambur A, Hall IW (1997) Dynamic compressive behavior of a SiCw/Al composite. Scripta Mater 37:193–198
17. Yang C, Zong YY, Zheng ZZ, Shan DB (2014) Experimental and theoretical investigation on the compressive behavior of aluminum borate whisker reinforced 2024Al composites. Mater Charact 96:84–92
18. Galiyev A, Kaibyshev R, Gottstein G (2001) Correlation of plastic deformation and dynamic recrystallization in magnesium alloy ZK60. Acta Mater 49:1199–1207
19. Senthilkumar V, Balaji A, Narayanasamy R (2012) Analysis of hot deformation behavior of Al 5083-TiC nanocomposite using constitutive and dynamic material models. Mater Des 37:102–110
20. Zener C, Hollomon JH (1944) Effect of strain rate upon plastic flow of steel. J Appl Phys 15:22–32
21. Sherby OD, Klundt RH, Miller AK (1977) Flow stress, subgrain size, and subgrain stability at elevated temperature. Metall Trans A 8A:843–850
22. Li Z, Samuel AM, Samuel FH, Ravindran C, Valtierra S (2003) Effect of alloying elements on the segregation and dissolution of $CuAl_2$ phase in Al-Si-Cu 319 alloys. J Mater Sci 38: 1205–1218
23. Zhou M, Clode MP (1998) Constitutive equations for modeling flow softening due to dynamic recovery and heat generation during plastic deformation. Mech Mater 27:63–76
24. Prasad YVRK, Gegel HL, Doraivelu SM, Malas JC, Morgan JT, Lark KA, Barker DR (1984) Modeling of dynamic material behavior in hot deformation: forging of Ti-6242. Metall Trans A 15A:1883–1892

Development and Characterization of In-situ Aluminum–Titanium Carbide Composites Prepared by Pneumatic Powder Injection Route

Sheetal Gupta, Anirban Giri, Saikat Adhikari and Vivek Srivastava

Abstract High costs, consistency and scalability are the major challenges in development of metal matrix composites. To overcome these challenges, a novel process has been demonstrated in which pneumatic powder injection was used instead of the conventional mechanical stirring process. In this study, aluminum–titanium carbide particulate composites were prepared in situ by injection of mixed salt of titanium (K_2TiF_6-potasium titanium fluoride salt) and graphite powder through submerged lance into molten aluminum. Uniform distribution of equiaxed Al_3Ti (aluminum–titanium intermetallic) particles and TiC (titanium carbide) particles were achieved depending on the reaction temperature and holding time. SEM-EDS confirmed the presence of submicron TiC particles distributed throughout the matrix. TiC particles generated in situ are thermodynamically more stable and tend to have cleaner matrix particle interfaces. More than 15% improvement in elastic modulus along with significant improvement in other mechanical properties was achieved by up to 10 wt% TiC reinforcement.

Keywords Metal matrix composites · In-situ · Pneumatic powder injection Modulus · TiC

Introduction

In recent times, aluminum alloys have been gaining widespread acceptance in the automotive industry [1]. Conventional wrought aluminum alloys, however, have limitations with maximum achievable modulus and fatigue, creep and wear resistance. As high-performance lightweight materials, aluminum matrix composites (AMCs) reinforced with harder and stiffer ceramic particles are widely used in

---

S. Gupta (✉) · A. Giri · S. Adhikari
Aditya Birla Science and Technology Company Private Limited, Navi Mumbai, India
e-mail: sheetal.gupta@adityabirla.com

V. Srivastava
Hindalco Industries Limited, Mumbai, India

T. S. Srivatsan et al. (eds.), *Metal-Matrix Composites Innovations, Advances and Applications*, The Minerals, Metals & Materials Series,
https://doi.org/10.1007/978-3-319-72853-7_5

aerospace, aircraft and automotive applications because of their excellent properties, such as high specific stiffness, high specific strength and excellent wear resistance [2]. Mass production of such materials of consistent quality and low cost, though, remains a challenge [3].

AMC manufacturing routes can be broadly classified as ex situ and in-situ. When the reinforcement is externally added to the matrix, ex situ composite materials are created. In situ synthesizing of metal matrix composites involves the production of reinforcements within the matrix during the fabrication process. Ex situ processes can be further sub-divided as either solid or liquid state process, though some gaseous state processes too have been reported in literature [4]. In brief, powder metallurgical (P/M) processes utilizes metal and reinforcement powders as the starting raw materials that are blended and consolidated to produce the composite. P/M route makes it possible to employ lower operating temperatures, overcome issues with particle wetting, minimize undesirable interfacial reactions and improve mechanical properties. Also, significantly higher reinforcement volume fractions can be incorporated through this route. P/M route, however, remains expensive compared to liquid metallurgy route, have higher levels of porosity and is suitable only for low volume manufacturing [5]. Liquid metallurgical processes for AMC synthesis include liquid infiltration, squeeze casting, spray co-deposition, stir casting and in-situ (reactive) processing.

The selection of the processing route depends on many factors including type and level of reinforcement loading and the degree of micro structural integrity desired [6]. For ex situ liquid metallurgical processes, lack of wetting of the reinforcement and interfacial reactions leading to poor interfacial bonding are major concerns. In-situ AMCs overcome these limitations but control of process variables remains an issue. Several liquid metallurgical process for in-situ production of particulate AMC have been reported in literature [7]. For these processes, kinetics of particle formation is not yet clear. The reaction rate depends on several factors like temperature, amount of stirring and presence of alloying elements etc.

There are several ceramic reinforcements available for Al-based metal matrix composites but in this study TiC has been considered due to its high hardness, stiffness and wear resistance properties [8–11]. These features of the TiC phase are combined in Al–TiC composites with the ductility, toughness, electrical and thermal conductivity of the aluminum matrix. Al–TiC composites can also potentially be used as grain refiner for aluminum alloys. TiC particles generated in-situ are also thermodynamically more stable, tend to have cleaner matrix–particle interfaces and are free from adsorbed gases, oxide films and detrimental reaction products. Aluminum matrix TiC composites have been reported earlier [2]. The following reaction sequence is reported for flux assisted synthesis of Al–TiC composite.

$$Al + K_2TiF_6 + C \rightarrow Al_3Ti + Al_4C_3 + KAlF_4 + KAlF_6 \ (750\,^{\circ}C) \quad (1)$$

$$Al_3Ti + Al_4C_3 \rightarrow Al + TiC \ (900\text{--}950\,^{\circ}C) \quad (2)$$

K_2TiF_6 and graphite react with molten aluminum at 750 °C (Reaction 1), where Ti is released from K_2TiF_6 and dissolved in the melt,forming Al_3Ti whereas graphite reacts with Al and forms Al_4C_3 particles. K-Al-F salt is also formed, which helps in cleaning the particle surfaces and remove the oxide layer from the surface of the melt. Al_3Ti and Al_4C_3 particles react above 900 °C to form TiC particles which are more stable than Al_3Ti particles (Reaction 2).

The present work was undertaken to synthesize Al–TiC composites in situ by reacting K_2TiF_6 and graphite in molten aluminum. The reactions involved in the synthesis were investigated by metallographic analysis of melt samples taken during the process.

Experimental Procedures

Particulate aluminum matrix composites (AMCs) from 0 to 10 wt% titanium carbide (TiC) reinforcement were prepared for this study using a modified salt flux assisted synthesis process.

Figure 1 shows the schematic experimental set up used for this study in which a 50 KW induction furnace with a 25 kg capacity silicon carbide crucible was employed. For powder injection, a powder hopper of 15 kg capacity, a venturi system and a lance assembly system was used. During synthesis, it was required to manipulate the lance assembly on (a) vertical axis to submerge the lance inside the molten metal and (b) horizontal axis during pouring the molten metal into the moulds. In order to facilitate these manipulations, the lance assembly was mounted on a screw-driven manipulator and a worm gear was placed between the steel channels of the mounting structure. A mating travelling gear moves on the vertical axis when the worm gear is operated. The lance assembly was attached to the travelling gear allowing the motion of the lance. This entire assembly itself was mounted on a turn table to enable movement of the lance assembly in horizontal direction through a swivel motion.

During synthesis, 15–20 kg commercial purity aluminum (99.5%) ingots were melted in an induction furnace at 800–900 °C. Graphite powder and K_2TiF_6 salts were premixed as required by stoichiometry and injected into the melt through specially designed alumina or graphite lance to avoid thermal shock and lance chocking. The flux and graphite powders were injected at room temperature and no preheating was required.

During pneumatic powder injection, argon or nitrogen was used as a carrier gas to agitate the melt and prevent lance chocking. After completing the injection of the powders, the melt was held for 30–60 min at the reaction temperature to allow the reaction to proceed to completion while continuing gas bubbling to ensure mixing. At the end of the hold time, spent salt and dross was skimmed and the composite was poured in the molds to cast MMC ingots.

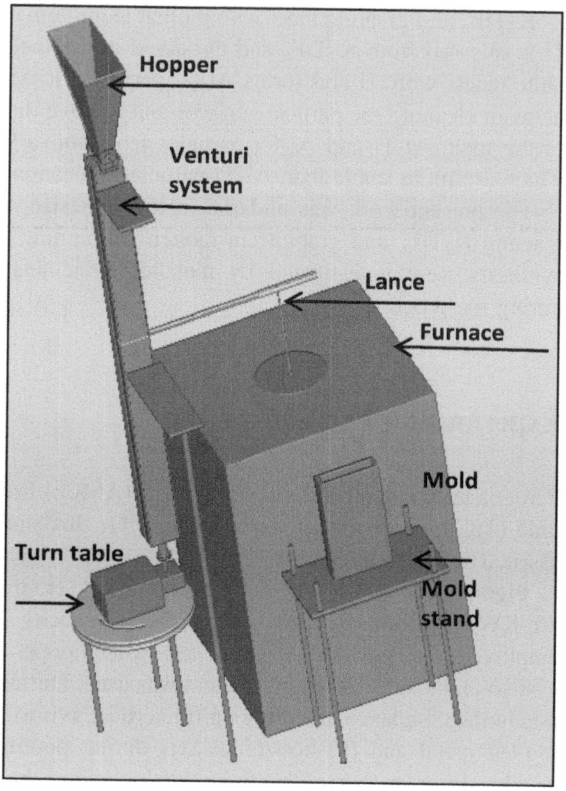

Fig. 1 Schematic illustration of the experimental set up used for metal matrix composite synthesis

The pouring temperature was controlled by the induction furnace to obtain different reinforcement morphology. Multiple samples were taken from the cast ingot for microstructure and hardness measurements. Vickers hardness of the specimens was measured at 5 kgf load with dwell time of 10 s. For measuring the tensile properties of the composite specimens, the cast samples were machined to standard tensile specimens according to ASTM: B557M-10 standard and tested. The stress–strain data were recorded by using video extensometer attached with the tensile testing machine (model Zwick-Z050) according to the load–displacement data.

Results and Discussion

Characterization of the cast composite samples was conducted to understand the formation of phases, their distribution within the composite and their effect on mechanical properties.

Optical Microstructure

Figure 2a, b show microstructures of 5 wt% TiC as-cast composites at low magnification. Figure 2a shows the presence of 20–50 μm sized blocky particles of Al_3Ti (confirmed by EDS) along with some residual graphite particles. Due to insufficient cleaning of the dross some of the carbon residue carried over from the melt which may affect the mechanical properties of the composite material. Figure 2b shows the SEM image of the 5 wt% TiC as-cast sample. Uniform distribution of blocky and needle shaped particles of Al_3Ti were observed throughout the matrix. TiC particles were not observed at lower magnifications. However, a distribution of sub-micron sized TiC particles was clearly identified at higher magnifications as seen in the SEM micrograph in Fig. 2c and EDS spectra in Fig. 2d. The clear majority of the Al_3Ti particles which formed at 750 °C were of blocky morphology as seen in Fig. 2a, b. Submicron TiC particles were formed at 900–950 °C after reacting of Al_3Ti particles with Al_4C_3 particles. This is in agreement with the reaction mechanism described earlier for in-situ formation of TiC in aluminum melts and reported in literature [2].

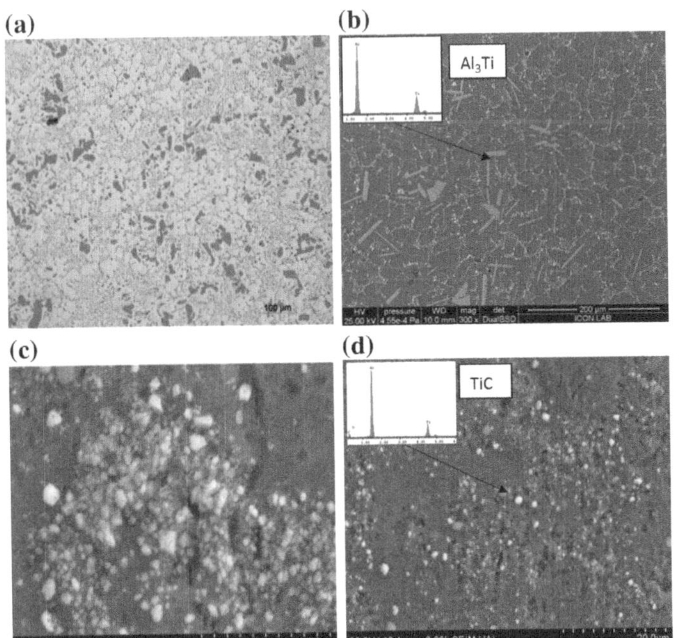

Fig. 2 Microstructure of Al-5 wt% TiC composites: **a** Low magnification image of the composites having large blocky particles of Al_3Ti and some residual graphite particles **b** SEM image of needles shape of Al_3Ti particles **c** Distribution of fine sub-micron sized TiC particles **d** SEM-EDS image of submicron sized TiC particles

X-Ray Diffraction

The microstructural investigations were confirmed by the XRD analysis, which revealed Al_3Ti and TiC reflections in aluminum matrix. Figure 3a shows the XRD pattern for two Al-5 wt%TiC composites prepared separately at 900 and 950 °C. The XRD pattern shows that maximum transformation of Al_3Ti to TiC could be achieved around 950 °C by the pneumatic powder injection process. Higher temperature favors the conversion of Al_3Ti particles to TiC particles. Figure 3b shows XRD pattern for different amount of reinforcement addition starting from 0 to 10 wt%, which clearly indicates increasing peak intensities of TiC with increasing amount of reinforcement addition.

Homogeneity and Consistency

During synthesis, homogeneity and consistency of the different phases in the matrix are very important to achieve desired mechanical and surface properties. These can

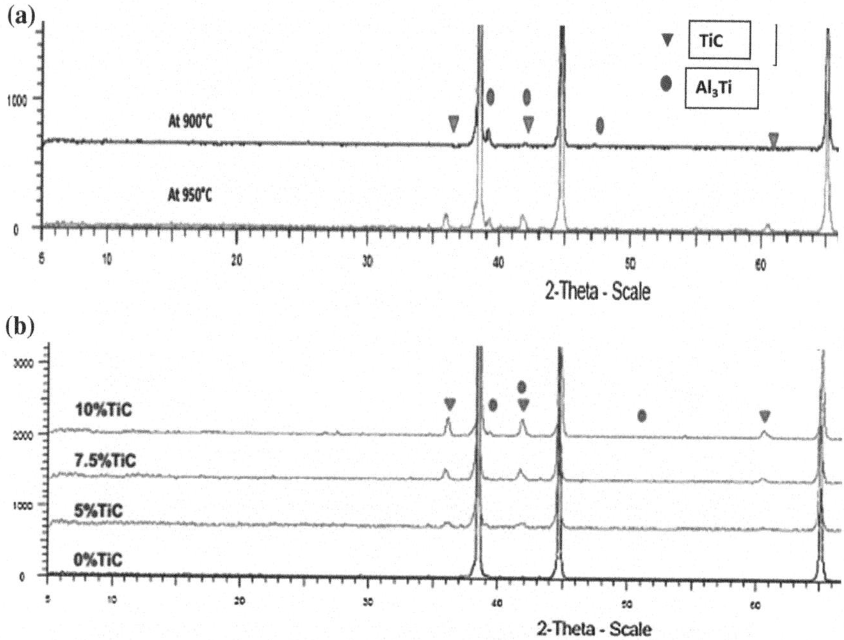

Fig. 3 XRD pattern for Al–TiC composites **a** XRD pattern showing formation of TiC at increasing reaction temperature **b** increasing peak intensities of reinforcing species for different amount of reinforcement addition

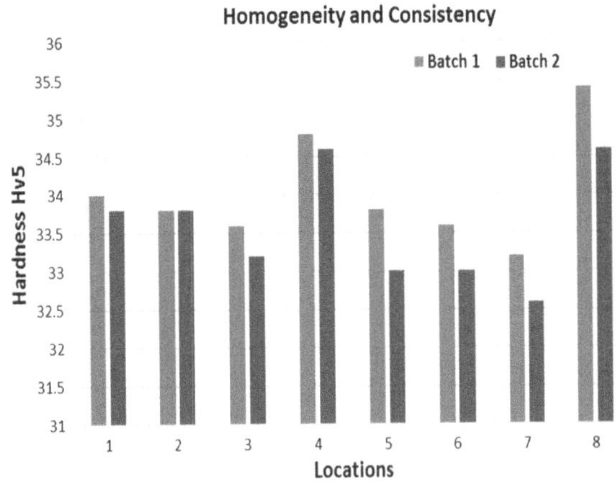

Fig. 4 Macro hardness of 5 wt% TiC composite at different locations and different batches

be analyzed by hardness measurement at different locations of the ingot from different batches. Figure 4 shows the effectiveness of the current synthesis route in consistently producing homogenous MMC material of 5 wt% TiC reinforcement. The homogeneity is achieved due to pneumatic injection of powder through a specially designed lance into the melt where argon or nitrogen were being used as a carrier gas. This process helps in agitating and mixing the powder uniformly throughout the melt to produce a fairly uniform macro-distribution of reinforcement particles which resulted in a ±3.0% variation in macro-hardness within one batch and only ±1.1% batch to batch variation.

Mechanical Properties

To understand the effect of hard TiC and Al_3Ti particles on overall mechanical properties, hardness, tensile and young's modulus properties were measured.

Macro-Hardness

Figure 5 shows the macro-hardness values for the MMC material with 0–10 wt% TiC reinforcement. All values of hardness are the average for ~5 samples from different batches. Figure 5 indicates that hardness values increase almost linearly with increasing TiC reinforcement. 5 wt% TiC reinforcement leads to 33% improvement in hardness over base alloy. Similarly, 7.5 and 10 wt% TiC reinforcement lead to 44 and 79% improvement in hardness respectively. These improvements in the hardness values are due to the addition of harder phases of TiC and Al_3Ti.

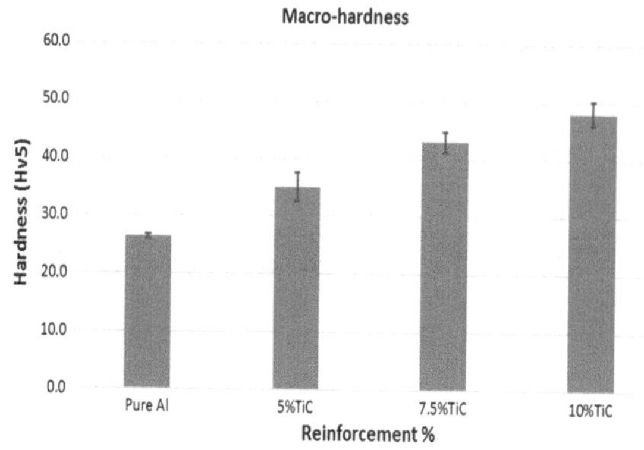

Fig. 5 Macro-hardness of 0–10 wt% TiC reinforcement composite materials

Tensile Properties

In the case of composites, reinforcements play a role in both the elastic and plastic regime. In the elastic regime, modulus of the composite is altered due to load transfer from the matrix to the composite. In the plastic regime, the deformation is primarily controlled by motion of dislocations in the matrix. Large reinforcement particles are mostly ineffective in restricting dislocation motion but do act as stress concentrators, thereby reducing the ductility of the composite. Finer particles, on the other hand, restrict dislocation motion, increase yield and tensile strength without severely compromising the ductility. In view of the above understanding, tensile properties were evaluated of 0–10 wt% TiC reinforcement composite materials.

The variation of yield strength, UTS and elongation values with TiC reinforcement are shown in Figs. 6, 7 and 8 respectively. Figures 6 and 7 clearly indicate that there is a marked increase in yield strength and UTS with increasing amount of TiC addition. Ductility of the composites as characterized by % elongation was found to be decrease with increasing reinforcement as seen in Fig. 8.

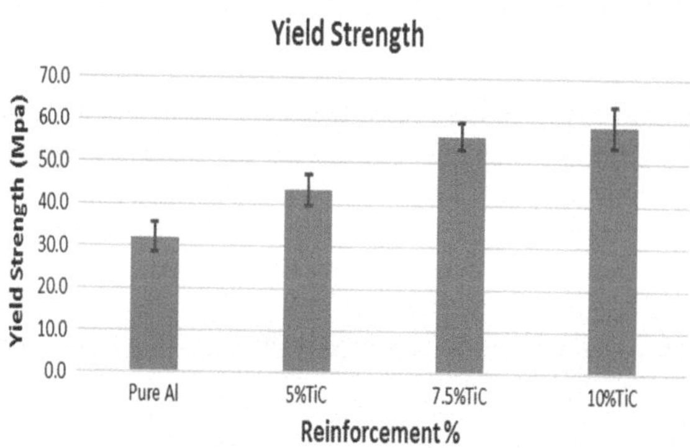

Fig. 6 Yield strength of 0–10 wt% TiC reinforcement composite materials

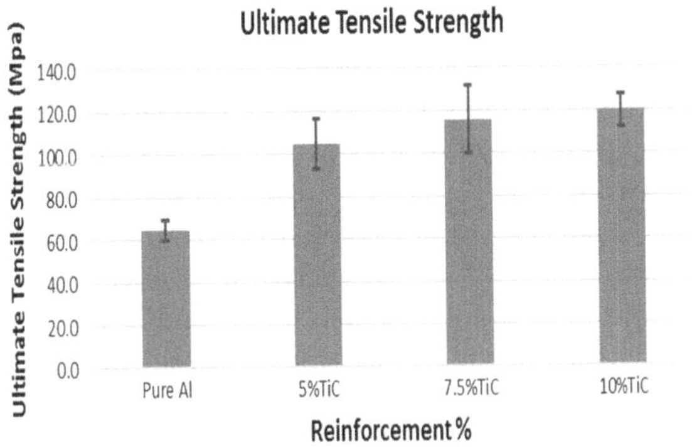

Fig. 7 Ultimate tensile strength of 0–10 wt% TiC reinforcement composite materials

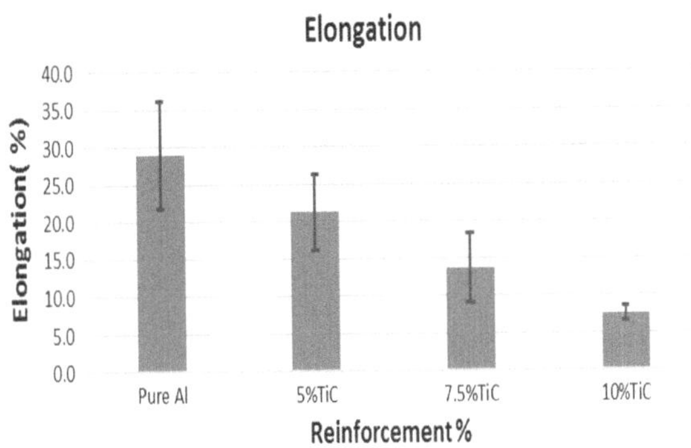

Fig. 8 Elongation of 0–10 wt% TiC reinforcement composite materials

Figure 9 shows dependence of elastic modulus of the composites with TiC reinforcement. The elastic modulus increases with an increasing amount of TiC addition. This is due to the reinforcements being able to partially carry the applied load and reduce the actual stresses in the matrix. This also demonstrates the superior interfacial bonding of the reinforcement to the matrix. There was about 15% improvement in observed for 10 wt% TiC reinforcement as compared to the base alloy.

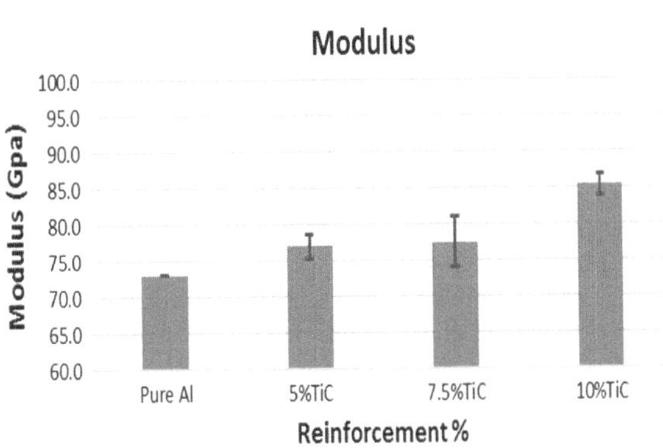

Fig. 9 Modulus of 0–10 wt% TiC reinforcement composite materials

Conclusions

1. At low reaction temperatures (750 °C), the composite contains only blocky Al_3Ti phases, which transforms to TiC as reaction temperature increases to 950 °C.
2. Volume fraction of Al_3Ti and TiC particles in Al–TiC composites increases with increasing K_2TiF_6 and graphite powder addition into the aluminum melt.
3. Macro-hardness variation within one batch was ±3.0% and batch to batch variation was ±1.1%. The novel pneumatic injection method helped in agitating and mixing the powder throughout the melt leading to homogenous properties in the composite material.
4. Mechanical properties like hardness, yield strength and tensile strength improved almost linearly with amount of TiC addition in the aluminum matrix.
5. Ductility of the composite showed linearly decreasing trend with increasing amount of TiC addition.
6. Young's modulus also improved significantly with amount of TiC addition. For 10 wt% TiC reinforcement composite materials, the young's modulus increased by 15% as compared to the base alloy.

References

1. Report on The road ahead-automotive materials. Ducker Worldwide Research Company 2016
2. Birol Y (2008) In situ synthesis of Al–TiC_p composites by reacting K_2TiF_6 and particulate graphite in molten aluminum. J Alloy Compd 454:110–117
3. Chawla KK (1998) Metal matrix composites. In: Composite materials. Springer, New York
4. Borgonovo C, Apelian D (2011) Manufacture of aluminum nanocomposites: a critical review. Mater Sci Forum 678:1–22
5. Liu YB, Lim SC, Lu L, Lai MO (1994) Recent development in the fabrication of metal matrix-particulate composites using powder metallurgy techniques. J Mater Sci 29:999–2007
6. Surappa MK (2003) Aluminium matrix composites: challenges and opportunities. In: Sadhana, vol 28, Parts 1 & 2. Department of Metallurgy, Indian Institute of Science, Bangalore, pp 319–334
7. Babalola PO, Bolu CA, Odunfa KM (2014) Development of aluminium matrix composites. Int J Eng Technol Res 2:1–11
8. Kadolkar PB, Watkins TR, De Hosson JTM, Kooi BJ, Dahotre NB (2007) State of residual stress in laser-deposited ceramic composite coatings on aluminum alloys. Acta Mater 55:1203
9. Vreeling JA, Ocelík V, Hamstra GA, Pei YT, De Hosson JTM (2000) In-situ microscopy investigation of failure mechanisms in Al/SiC_p metal matrix composite produced by laser embedding. Scripta Mater 42:589
10. Anandkumar R, Almeida A, Colaço R, Vilar R, Ocelik V, De Hosson JTM (2007) Microstructure and wear studies of laser clad Al-Si/$SiC_{(p)}$ composite coatings. Surf Coat Technol 201:9497
11. Torres B, Lieblich M, Ibáñez J, García-Escorial A (2002) Mechanical properties of some PM aluminide and silicide reinforced 2124 aluminium matrix composites. Scripta Mater 47:45

Part III
Basic History and Advances in Metal Matrix Composites

Microstructure and Mechanical Behavior of Cryomilled Al–Mg Composites Reinforced with Nanometric Yttria Partially Stabilized Zirconia

Matthew Dussing, Hanry Yang, Troy D. Topping, Enrique J. Lavernia, Kaka Ma and Julie M. Schoenung

Abstract The present work investigated the viability of using nanometric 3 mol% yttria partially stabilized zirconia (3YSZ) as a particulate reinforcement in ultrafine grained aluminum alloy matrix composites. Four types of composite materials, with variable amounts of coarse grain regions and volume fractions of 3YSZ, were fabricated through cryomilling and hot isostatic pressing; one of the materials was extruded. Al–Mg alloy (AA5083) was selected as the matrix alloy. Microstructural characterization revealed that the 3YSZ particles were well dispersed in the Al matrix. The grain sizes of the Al matrix ranged from 77 ± 41 to 362 ± 185 nm depending on the thermomechanical processing. The composite with 2.2 vol% 3YSZ exhibited an ultimate tensile strength of 795 MPa and a strain-to-failure of 1.84%. In contrast, the composite with 10 vol% 3YSZ exhibited an ultimate compressive strength of 611 MPa with 22.5% strain-to-failure; this composite also retained 30% of its room temperature strength at 673 K (400 °C). Evaluation of strengthening mechanisms suggests that Hall-Petch strengthening is a predominant mechanism controlling the achieved strength.

M. Dussing
Department of Chemical Engineering and Materials Science, University of California Davis, Davis, CA 95616, USA

H. Yang
Keysight Technologies, Santa Rosa, CA 95403, USA

T. D. Topping
Department of Mechanical Engineering, California State University, Sacramento, Sacramento, CA 95819, USA

E. J. Lavernia · J. M. Schoenung (✉)
Department of Chemical Engineering and Materials Science, University of California, Irvine, Irvine, CA 92617, USA
e-mail: julie.schoenung@uci.edu

K. Ma
Department of Mechanical Engineering, Colorado State University, Fort Collins, CO 80523, USA

T. S. Srivatsan et al. (eds.), *Metal-Matrix Composites Innovations, Advances and Applications*, The Minerals, Metals & Materials Series,
https://doi.org/10.1007/978-3-319-72853-7_6

Keywords Al alloy matrix composites · Ultrafine grained structure
Yttria-stabilized zirconia · Mechanical behavior

Introduction

Aluminum-based metal matrix composites (MMCs) have attracted extensive research interest in recent decades because of the broad spectrum of property combinations that can be achieved, such as a combination of high specific strength and stiffness, compared to unreinforced Al alloys [1–5]. The microstructure and mechanical properties of the Al MMCs depend on the selection of the matrix alloy, reinforcement, and the processing methods [1, 5]. Cryomilling has been found to be a cost-effective method of manufacturing ultrafine-grained (UFG) metals, alloys, and MMCs [6–9]. Published studies on cryomilled MMCs report well dispersed reinforcement particles with clean interfaces [8, 10]. Previous work on cryomilled MMCs has focused on an Al alloy 5083 (AA 5083; Al–4.4Mg–0.7Mn–0.15Cr wt%) matrix with micron, submicron, and nanometric sized B_4C reinforcement [6, 11–16]. Research results indicate that it is possible to increase strength, sometimes at lower reinforcement volume percentages, by moving to nanometric reinforcements [3, 17, 18].

Zirconia (ZrO_2) is a ceramic that has been intensively studied for nearly 40 years, partly due to its high fracture toughness for a ceramic [19]. For example, bulk Al_2O_3 and SiC have a fracture toughness of 4 MPa $\sqrt{m}$ while 3YSZ exhibits a fracture toughness of 10 MPa $\sqrt{m}$ [20]. The high fracture toughness of 3YSZ is attributed to the transformation from a metastable tetragonal phase to a monoclinic phase [19]. Small amounts of various oxide dopants, such as Y_2O_3, MgO, and CaO, allow the high-temperature tetragonal phase to remain metastable at room temperature and below [21]. The highest combination of fracture toughness and strength has been found with 3 mol% Y_2O_3 stabilization [22]. Previous work has shown impressive results by adding stabilized zirconia to Al_2O_3, creating zirconia toughened alumina [23–27]. Toughness values increased from 4.2 MPa $\sqrt{m}$ in pure Al_2O_3 to 7 MPa $\sqrt{m}$ with 20 vol% 3YSZ reinforcement [24, 25].

Particle and grain sizes in these studies are on the order of 200 nm for the 3YSZ and 2 μm for the α-Al_2O_3, which is approximately comparable to the material created in this research with an Al matrix [24]. Only a few studies have been published describing the use of zirconia as an Al-based metal matrix nanocomposite (MMNC) reinforcement [28, 29]. Geng et al. simply mixed 5 and 10 vol% zirconia powder with coarse grained (CG) pure Al, and found tensile strength increased from 34.5 to 72.3 MPa, with a corresponding reduction of ductility from 18.6 to 3.2% [28]. However, this material was processed at temperatures approaching 600 °C which resulted in significant reinforcement-matrix reactions, formation of Al_3Zr, and retard the transformation toughening. Dutkiewicz et al. ball milled 10 vol% 3YSZ with Al–Mg–Si alloy (AA6061) powder in an argon atmosphere, and consolidated the ball milled composite powder to bulk via vacuum hot pressing. The bulk Al composite exhibited a compressive strength of 1000 MPa with a 7% strain-to-failure [29].

In view of the above discussion, the objective of this work is to investigate the feasibility of using nanometric 3 mol% yttria-stabilized zirconia (3YSZ) as the reinforcement in a cryomilled AA 5083 matrix. Relevant research questions include the following: Is the 3YSZ homogeneously distributed in the cryomilled powder? What is the impact of 3YSZ on mechanical properties? To answer these questions, the microstructure of four types of composite materials, with variable amounts of coarse grain regions and volume fractions of 3YSZ, were characterized in detail. Mechanical performance was studied though hardness, compression at room temperature and elevated temperature, and tension testing. The strengthening mechanisms are evaluated.

Materials and Experimental Procedures

Gas-atomized-325 mesh (particle diameter < 44 μm) AA 5083 powder manufactured by Valimet, Inc. (Stockton, CA) was used as the primary feedstock material. Two 3YSZ reinforcement powders were used: Material A utilized TZ-3Y-E powder from Tosoh Corp. (Tokyo, Japan) and Materials B and C utilized HSY-3F powder from Daiichi Kigenso Kagaku Kogyo Co., Ltd. (Osaka, Japan). Both powders are 3 mol% yttria partially stabilized zirconia and are supplied as easily dispersed spherical agglomerates. The TZ-3Y-E powder had an average particle size of 77 ± 16 nm, and the HSY-3F powder had an average particle size of 104 ± 29 nm. Pre-mixed AA 5083 and 3YSZ powder was cryomilled in a modified 1-S Szegvari Attritor (Union Process, Akron, OH) in liquid nitrogen. Additional information on cryomilling is available from Witkin and Lavernia [30]. Cryomilling time was reduced from 12 h for Material A to 8 h for Materials B and C. This change was intended to create larger grain sizes in the UFG region in the final bulk sample and thus to improve the ductility [31, 32]. 0.2 wt% stearic acid ($CH_3(CH_2)_{16}CO_2H$) was added to reduce excessive cold welding of the powder during cryomilling [33].

After cryomilling, AA 6061 cans were filled with powder and hot vacuum degassed to remove hydrogen introduced by stearic acid as well as chemi- and physisorbed H_2O from the powder surface [33]. Material A was degassed at 773 K (500 °C) for 12 h, while Materials B and C were degassed at a lower temperature of 688 K (415 °C) for 30 h to avoid chemical reactions between the Al matrix and the 3YSZ particles. A final pressure of 10^{-6} torr was achieved during degassing. Details of materials composition, cryomilling and degassing parameters, and chemical content are provided in Table 1. Gas atomized AA5083 powder was blended with cryomilled powder prior to degassing and consolidation to create CG regions (nominally 30 wt%) in Material C.

Table 1 Material composition, cryomilling and degassing parameters, and chemical content

Materials	A, A+Ext	B	C
Composition (wt%)	95% UFG AA 5083 5% YSZ	80% UFG AA 5083 20% YSZ	55.9% UFG AA 5083 14.1% YSZ 30% CG AA 5083
Reinforcement (vol%)	2.2	10	7
Reinforcement diameter (nm)	77 ± 16	104 ± 29	104 ± 29
Cryomilling time (h)	12	8	8
Degassing temperature	773 K (500 °C)	688 K (415 °C)	688 K (415 °C)
Degassing time (h)	12	30	30
H content (ppm)	37	387	386
N content (%)	0.870	0.894	0.610

Note Nominal composition of AA5083 is Al–4.5Mg–0.57Mn–0.24Fe (by weight percentage)

Separate degassed cans of Materials A, B, and C were hot isostatic pressed (HIPed) in a 172 MPa (25 ksi) argon atmosphere at 673 K (400 °C) for 4 h in a Flow Autoclave Systems Inc. (Columbus, OH) HIP. Following HIPing, the can material was machined off of the billet. As-HIPed samples were sectioned from the billet by electrical discharge machining (EDM) using a model FA20S from Mitsubishi (Tokyo, Japan). For Material A, the remaining billet was slow strain rate extruded into a 5/16 in (7.9 mm) diameter cylindrical rod using a 10:1 area reduction ratio after a preheat of 1 h at 673 K (400 °C) in a Hypress Technologies, Inc. (Oxnard, CA) press. This extruded material is referred to as Material A+Ext. Extrusion attempts on Materials B and C were unsuccessful. A thermal exposure profile of the processing steps is provided in Fig. 1.

Fig. 1 Thermal exposure of materials from cryomilling through final thermomechanical processing

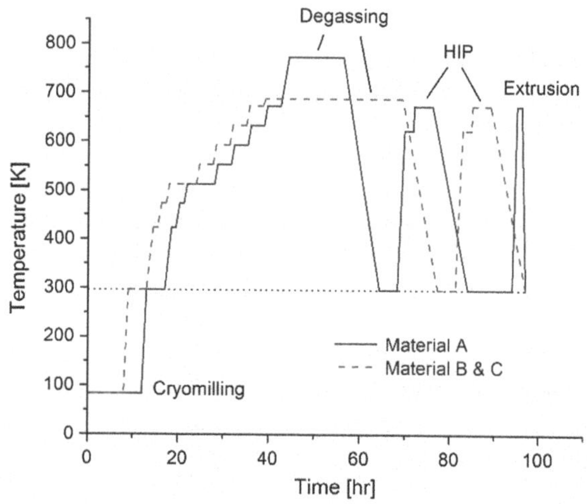

H and N concentrations were measured by LECO Corp. (St. Joseph, MI) using inert gas fusion analyzers for H (RH-404) and N (TCH-600). Cross-sections for SEM were argon ion beam polished in a JEOL (Peabody, MA) SM-09010 cross-section polisher. Ion polishing was chosen over traditional mechanical polishing to preserve porosity and other material characteristics that would be obscured by traditional polishing. An FEI (Hillsboro, OR) 430 NanoSEM operating at 5 kV was used for general microstructural characterization. Electron backscatter diffraction (EBSD) mapping of CG regions was performed on an FEI XL30 SEM operating at 15 kV fitted with an Oxford Instruments (Concord, MA) EBSD detector and software suite. CG vol% was estimated from the CG area %. Samples for EBSD mapping were hand polished to a mirror finish with 0.05 μm colloidal silica suspension. Samples for transmission electron microscopy (TEM) characterization were prepared by mechanically thinning with SiC paper to a thickness of approximately 60 μm. From this foil, 3 mm diameter discs were punched and jet polished to electron transparency using a 25% nitric acid and 75% methanol solution at 15 V at 233 K (−40 °C) in a Struers (Westlake, OH) Tenupol-3 jet polisher. TEM imaging was performed on a Philips (Andover, MA) CM 12 TEM operating at 120 kV. Energy-dispersive X-ray spectroscopy (EDS) using an Oxford detector was performed on the CM 12 as well. UFG grain size was calculated by measuring the maximum diameter of 200 grains for each material from TEM images. A grain size cutoff of 750 nm was chosen to differentiate between UFG and CG.

Density was measured on bulk samples via the standard Archimedes method. Hardness testing was performed on flat bulk samples polished with 1200 grit SiC paper, on a Buehler (Lake Bluff, IL) MicroMet 2004 Vickers microhardness testing machine with a 500-g load. Compression testing was performed on 3.5 mm by 3.5 mm by 4.5 mm HIPed cuboids of Materials A, B, and C polished to 1200 grit on all sides. High temperature (673 K, 400 °C) testing recorded machine displacement to estimate strain. Tensile testing was performed on ASTM E8 standard tensile specimens for Material A+Ext. These dog bone-shaped specimens were 0.25 in (6.35 mm) in diameter, 1.5 in (38.1 mm) long, and were machined from the extruded rod on a lathe. All compression and tension testing was performed on an Instron (Canton, MA) 8801 mechanical testing system at a strain rate of 10^{-3} s^{-1}.

Results and Discussion

Microstructure

The microstructural characteristics of the four bulk materials (A, A+Ext, B, C) were investigated using SEM, EBSD, and TEM. Figure 2 reveals the heterogeneous microstructure of Material A. Both CG regions (free of second phase particles) and

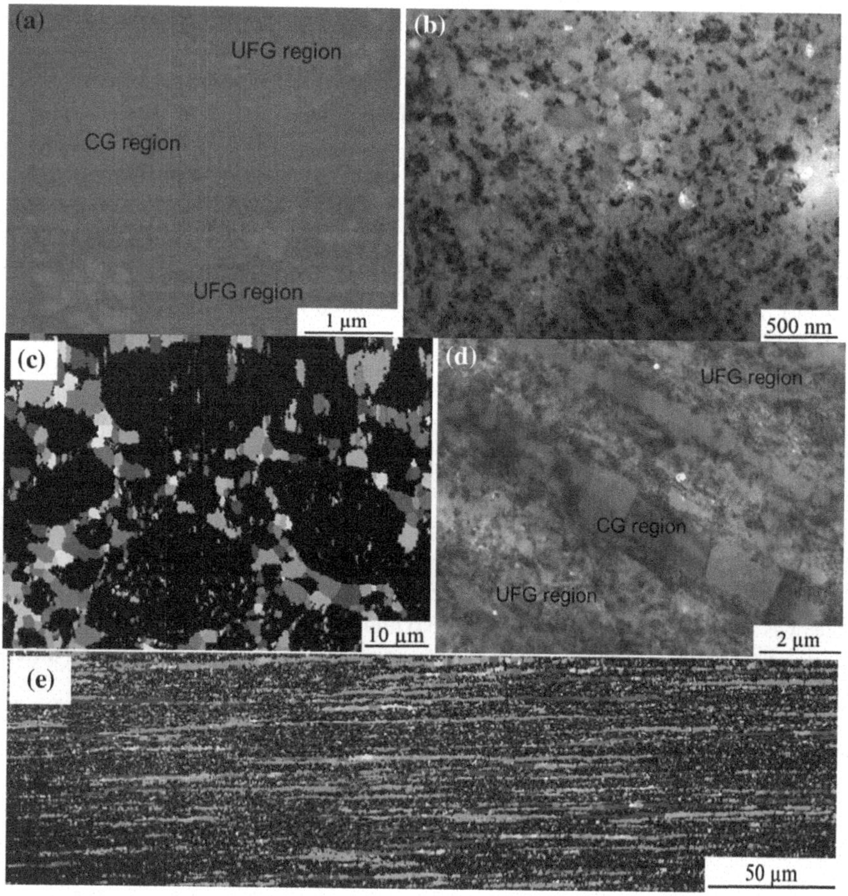

Fig. 2 Representative micrographs of **a** SEM of Material A showing UFG and CG regions, **b** TEM of Material A—UFG region, **c** EBSD of Material A indicating extensive CG regions (indexed and colored regions); the black regions are UFG regions that were not indexed in EBSD due to severe plastic deformation, **d** TEM of Material A+Ext showing CG regions drawn into bands, **e** EBSD of Material A+Ext with high concentration of CG bands

UFG regions (rich in second phase particles) are observed in Material A (Fig. 2a) [34]. The large white particles, approximately 200–400 nm in length, are likely Al_3Zr, Al_2O_3, or intermetallics such as Mg_2Si, Mg_2Al_3 or $Al_6(Mn,Fe)$, as described in other investigations [35, 36]. The UFG regions of A are shown in the TEM micrograph (Fig. 2b). The average grain size in the UFG region in A is 77 ± 41 nm with a grain aspect ratio of 1:6. The CG regions are present at the prior particle boundaries (PPBs) between the cryomilled powder particles, as shown in Fig. 2c [34]. The average grain size in the CG region in A is approximately

1.51 ± 0.74 μm. The UFG regions have low EBSD indexing, reflected as the black regions in Fig. 2c, due to high strain imparted on the material during cryomilling, and the limited spatial resolution of EBSD. Representative images of the microstructure in A+Ext are shown in Fig. 2d, e. In comparison to the as-HIPed Material A, the CG regions in A+Ext exhibits an elongated band structure. This is a result of the extrusion with a 10:1 area reduction ratio. The elongated bands extended to a few hundreds of μm in length, which consist of several grains with low angle boundaries. The average grain size in the CG region in A+Ext is 1.24 ± 0.77 μm with an aspect ratio up to 40:1. The average grain size in the UFG region in A+Ext is 362 ± 185 nm with a grain aspect ratio of 1:9. The UFG regions in A+Ext exhibit significant strain induced grain growth resulting from the secondary processing (i.e., extrusion) compared to those in A. This is consistent with recently published results suggesting that the presence of shear strain during deformation of UFGs facilitates grain rotation and migration, and thereby grain growth [37].

The microstructure of Material B is shown in Fig. 3. Material B contains a higher volume fraction of 3YSZ particles (10%) than A (2.2%), which is clearly indicated by the comparison between Figs. 2a and 3a. The 3YSZ particles are present as the fine white particles in Fig. 3a in the UFG region. Despite the high density, the 3YSZ particles are well dispersed with no apparent agglomeration. This

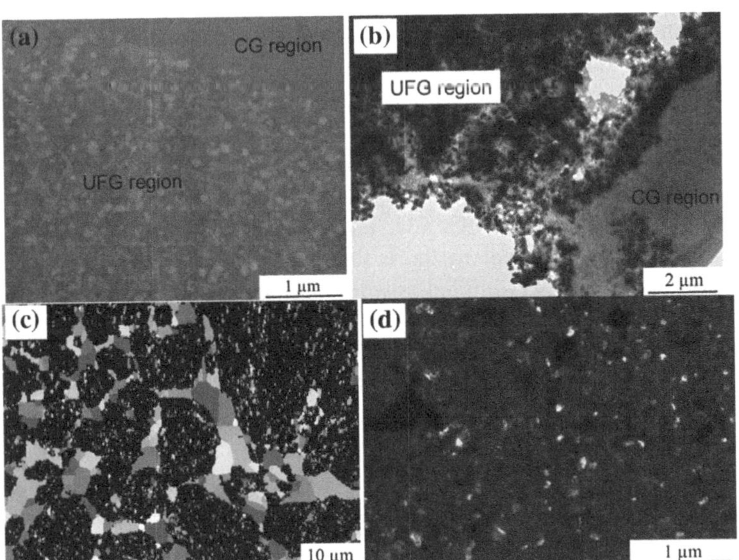

Fig. 3 Representative micrographs of: **a** SEM of Material B showing UFG and CG regions **b** TEM of Material B **c** EBSD of Material B indicating extensive diffused CG regions **d** Dark-field TEM of Material B (Al grains diffracting) illustrating technique to differentiate between YSZ and Al grains

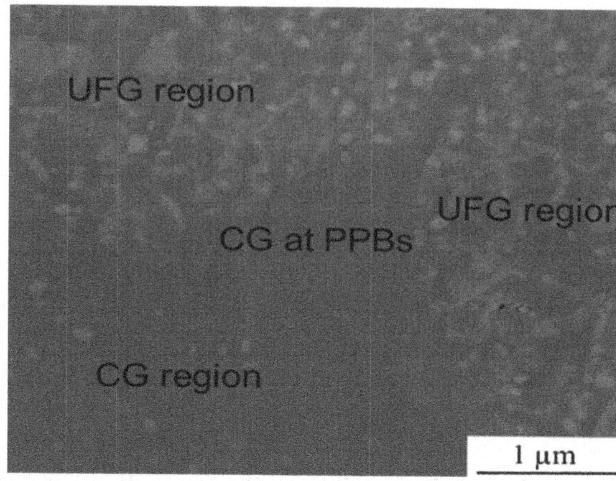

Fig. 4 Representative SEM micrograph of Material C

is important in the manufacture of metal matrix nanocomposites as nanometric reinforcements tend to agglomerate heavily during conventional processing methods [3]. Cryomilling involves both the matrix and reinforcement in their solid phases, and thus the wetting angles between matrix and reinforcement are not of concern. No evidence of poor bonding, such as pores or microcracks, was observed at the interface between the 3YSZ and the Al matrix. Most dark particles in the TEM micrograph (Fig. 3b) are 3YSZ, with the remainder being other intermetallic particles from the Al alloying elements. Material B also contains CG regions, as seen on the right side of Fig. 3b. The grain size of the CG region was determined through EBSD (Fig. 3c). The average grain size in the CG region in B is 1.54 ± 0.85 μm. The average grain size in the UFG region in B is approximately 108 ± 51 nm with a grain aspect ratio of 1:65.

Figure 4 shows the microstructure of Material C, which illustrates a junction of all three microconstituents: UFG, CG, and spherical 3YSZ particles. The UFG regions in C appear identical to those in B, as it is expected given that both materials used the same cryomilled powder. Gas atomized powder was added to create CG regions and consequently a trimodal microstructure in C [5, 8, 11, 14]. The CG regions (bottom-left of Fig. 4) can be differentiated by the intermetallic particles characteristic of CG AA 5083 [34, 36]. The CG region between the PPBs of the cryomilled powder is relatively free of intermetallic particles [34]. The UFG region has approximately the same concentration of 3YSZ as Material B.

Figures 5a–f present histograms of the UFG and CG grain sizes for each material. The characteristics of Materials A, A+Ext and B, such as the grain size and aspect ratio in CG and UFG regions, are summarized in Table 2. Needle-like dispersoids were observed in Material A, as shown in Fig. 6. Material A was degassed at 773 K (500 °C), which may lead to the reaction between Al and 3YSZ to form Al_3Zr intermetallics, as suggested by preliminary DSC data, not presented here [28].

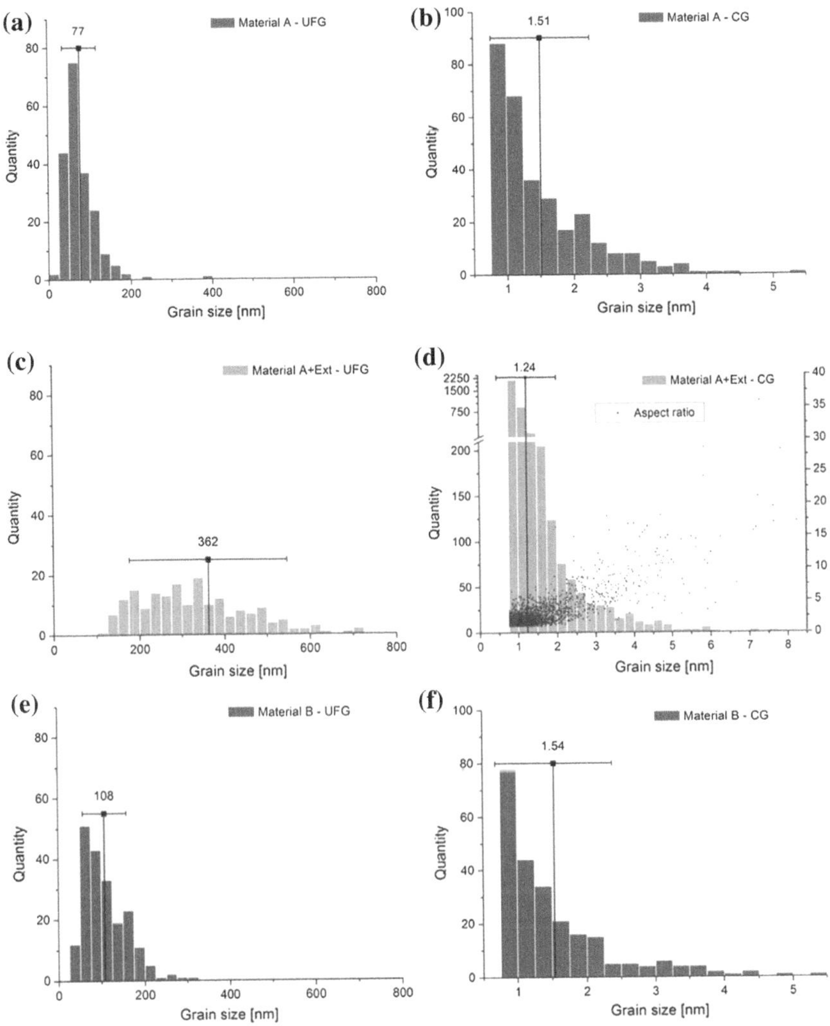

Fig. 5 Grain size histograms of **a** UFG in A, **b** CG in A, **c** UFG in A+Ext, **d** CG in A+Ext, **e** UFG in B, and **f** CG in B

Table 2 Grain sizes and aspect ratios of Materials A, A+Ext and B

Material	A	A+Ext	B
UFG size (nm)	77 ± 41	362 ± 185	108 ± 51
UFG aspect ratio	1.59 ± 0.30	1.93 ± 0.63	1.65 ± 0.35
CG size (μm)	1.51 ± 0.74	1.24 ± 0.77	1.54 ± 0.85
CG aspect ratio	1.96 ± 0.80	Up to 40:1	1.81 ± 0.61
Diffused CG (vol%)	24.0	27.4	22.8

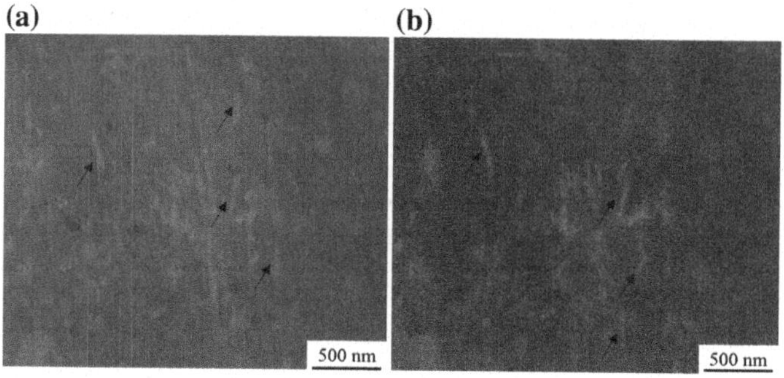

Fig. 6 **a** Secondary electron (SE) image, and **b** Backscattered electron (BSE)image of Material A+Ext showing the presence of needle-like dispersions, highlighted by red arrows

Mechanical Behavior

The density and mechanical testing results are summarized in Table 3. Material A exhibited a hardness of 209 HV, whereas C has a hardness of 123 HV due to the addition of 30 wt% of CG. Materials A, B, and C were compressively tested at both room temperature (RT) and 673 K (400 °C). The testing at elevated temperature (673 K) was intentionally stopped at 30% strain to failure before the samples fractured. Representative compressive stress-strain curves are shown in Fig. 7. At room temperature, Material A exhibited both higher yield strength and ultimate strength (610 and 828 MPa, respectively) than B and C; while the compressive strain to failure of A is lower than that of B and C. When tested at elevated temperatures, Material A retained 16% of its RT YS and 13% of its RT ultimate strength, whereas Material B retained 30% of its RT strength and Material C retained 29% of its RT strength. The YS (160 MPa) and ultimate strength (182 MPa) of Material B are higher than previously reported elevated temperature

Table 3 Summary of density and mechanical testing results

Material	A	A+Ext	B	C
Density (%)	98	97	92	95
Hardness (HV)	209 ± 4.8	179 ± 5.2	170 ± 5.6	123 ± 7.5
Test type	Compression	Tension	Compression	Compression
296 K yield strength (MPa)	610	690	526	379
296 K ultimate strength (MPa)	828	795	611	430
296 K strain to failure (%)	6.26	1.84	22.5	18.7
673 K yield strength (MPa)	95	–	160	110
673 K ultimate strength (MPa)	110	–	182	125
673 K strain to failure (%)	>30	–	>30	>30

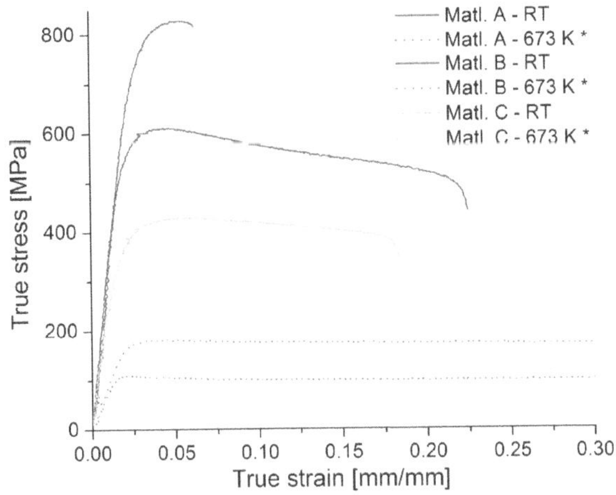

Fig. 7 Compressive stress-strain curves for Materials A, B, and C at room temperature and at 673 K (400 °C). *Strain data for 673 K (400 °C) measurements are estimated from load frame arm displacement

results. A UFG AA 5083 composite reinforced with 5 vol% nanometric B_4C particles exhibited a YS of 60 MPa at 673 K, despite having a compressive yield strength as high as 959 MPa at RT [38]. Ye et al. also reported similar results with a high-strain rate extruded AA 5083 composite with 10 vol% of 1–7 μm B_4C reinforcement material: the ultimate compressive strength was only 90 MPa at 673 K although its RT strength was approximately 1100 MPa [39]. Both Materials B and C were stronger than the two composites reported in [38, 39] at 673 K, despite lower room temperature strengths. The selection of the chemistry of the reinforcement affects the mechanical properties at elevated temperatures. The formation of Al–B–C phases due to the reaction between B_4C and Al significantly deteriorated the mechanical properties of B_4C/Al-derived composites [7]. In contrast, the formation of Al_3Zr can potentially strengthen the composite at elevated temperatures [40–44].

A representative tensile stress-strain curve for Material A+Ext is presented in Fig. 8. For comparison purposes, tensile testing was also performed on unreinforced AA 5083 alloys that were fabricated through the same thermomechanical processing route as A+Ext: one was made from gas atomized powder, while the other was made from cryomilled powder. A+Ext exhibited significantly higher strength (YS ~690 MPa and UTS ~795 MPa) than the two unreinforced alloys. However, the ductility of A+Ext is low (1.84%). Tensile testing was not conducted on the as-HIPed materials (A, B, C), because without extrusion, the PPBs limit ductility, as reported in prior work [45].

Strengthening Mechanisms

The strength of metal matrix composites is attributed to both the matrix and the reinforcement, as well as the interface between them. This section quantitatively

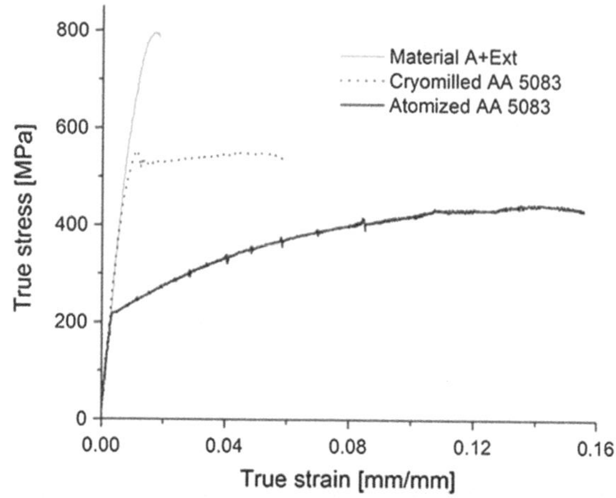

Fig. 8 Tensile stress-strain curves of Material A+Ext and two counterpart materials: unreinforced UFG AA5083 (cryomilled, degassed, HIPed and extruded), and CG AA 5083 made from gas atomized powder (degassed, HIPed and extruded)

estimates the strengthening contributions from different components in the 3YSZ/Al composites at room temperature. For a nanostructured or UFG Al–Mg alloy matrix the primary strengthening contributions come from grain size refinement (Hall-Petch) and solid solution strengthening. In materials that have been processed by cryomilling, which is a form of severe plastic deformation, a high density of dislocations form, then transition into dislocation cells and eventually convert to high angle grain boundaries (HAGBs) [30]. These HAGBs restrict dislocation motion and strengthen the materials through the Hall-Petch effect [46, 47]. Solid solution strengthening results from substitutional alloying elements, Mg in this case, distorting the Al lattice spacing. Here the Hall-Petch constant, k_{HP}, is modified to account for solid solution strengthening [38]. The Hall-Petch strengthening can be estimated by [46, 47]:

$$\sigma_{HP} = k_{HP} \Big/ \sqrt{d} \tag{1}$$

where d is the average grain diameter, k_{HP} is the Hall-Petch constant and equals 0.165 for AA5083 alloy [38]. Using the grain sizes listed in Table 2, the Hall-Petch strengthening is estimated to be 595, 274, and 502 MPa for Materials A, A+Ext, and B, respectively.

All the materials in this work contain a significant volume fraction of CG regions. The grain boundaries between CG grains are primarily low-angle grain boundaries after thermomechanical processing, which provide less resistance to dislocation motion [7, 12]. The contribution of CG grain boundaries to strength has been estimated by Hansen to be [7, 12, 48]:

$$\sigma_{CG} = M_T \alpha G \sqrt{1.5 b S_V \theta} \tag{2}$$

where $M_T = 0.3$ is the Taylor factor, $\alpha = 1$ is a constant, $G = 26.4$ GPa is the shear modulus, $b = 0.286$ nm is the Burgers vector, $S_V = 6/d$ is an approximation for grain boundary area per unit volume, and $\theta = 10°$ (in rad) is the average misorientation angle [49]. The strengthening from CG were calculated to be 137, 151, and 135 MPa for Materials A, A+Ext, and B, respectively.

Orowan strengthening is not considered to be an operable mechanism for the current study, because the size of the 3YSZ particles are almost the same as the Al grains. For Material A the 3YSZ particles are 77 ± 16 nm with Al grains of 77 ± 41 nm, and for Material B the 3YSZ particles are 104 ± 29 nm with Al grains of 108 ± 51 nm. Instead, load transfer from the Al matrix to the 3YSZ particles is considered. The strengthening contribution is estimated using the shear-lag model [11, 12]:

$$\sigma_{LT} = f_R \sigma_R \tag{3}$$

where f_R is the volume fraction of the reinforcement, and σ_R is the strength of the reinforcement. For this work $f_R = 0.022$ for Materials A and A+Ext, $f_R = 0.1$ for Material B, and $\sigma_R = 1500$ MPa [50] for 3YSZ. Load transfer strengthening values are thus calculated to be 33 MPa for Materials A and A+Ext, and 150 MPa for Material B.

A summary of the estimated values for these three different strengthening mechanisms is provided in Table 4. Comparing these values, it is apparent that the Hall Petch strengthening is dominant. Additional work to quantify the effect of the dislocations generated from elastic modulus mismatch and thermal expansion mismatch between the 3YSZ and the Al matrix is needed to comprehensively investigate the strengthening mechanisms. In addition, previous work has reported that adding Zr and Sc to Al–Mg alloys greatly increases their strength due to $Al_3(Sc, Zr)$ precipitation [40–42]. Thus, quantification of the effect of Al_3Zr is also needed, especially for Materials A and A+Ext. The Orowan strengthening from Al_3Zr may compensate for the loss in Hall-Petch strengthening due to the increased grain size in the UFG region in Material A+Ext.

Table 4 Contributions of strengthening mechanisms

Material	σ_{HP}	Vol% of UFG	σ_{CG}	Vol% of CG	σ_R	Vol% of 3YSZ	Experimental YS
A	595	73.8	137	24.0	33	2.2	610
A+Ext	274	70.6	151	27.4	33	2.2	690
B	502	67.2	135	22.8	150	10.0	526

σ in MPa

Summary

The present work demonstrated the feasibility of using nanometric 3YSZ particles as a reinforcement in UFG Al-based alloy composites. Various compositions and processing parameters were utilized to understand their impacts on the characteristics of the consolidated bulk materials. Microstructure and mechanical behavior were characterized, including elevated temperature (673 K, 400 °C) compression testing. Key findings are summarized as follows:

- 3YSZ particles were homogeneously distributed after cryomilling with no apparent agglomeration. Interfaces between reinforcement and matrix were clean and free of voids.
- As-HIP'ed Al nanocomposites with 2.2 vol% 3YSZ exhibited the highest compressive strength (828 MPa) with a relatively lower strain to failure than the materials that were cryomilled for a shorter time and reinforced with 10 vol% 3YSZ. The extruded Al nanocomposites exhibited significantly higher tensile strength (795 MPa) than the unreinforced Al alloy counterparts, despite limited tensile strain to failure (1.84%).
- Analysis of strengthening mechanisms revealed that Hall-Petch is the predominant mechanism. However, additional work is necessary to study the strengthening effect from dislocations and Al_3Zr.
- As-HIP'ed Al nanocomposites with 10 vol% 3YSZ exhibited superior compressive strength at 673 K compared to other literature data.

These results suggest that Al-3YSZ MMNCs are promising materials that merit further investigation, especially for improved mechanical performance at elevated temperatures. Because the microstructure evolution is sensitive to processing parameters and temperature, their properties can be tailored by carefully controlling the thermomechanical processing routes.

Acknowledgements Financial support from the Office of Naval Research (Grant Nos. N00014-12-1-0237 and N00014-12-C-0241) is gratefully acknowledged. Additionally, the authors would like to thank Daiichi Kigenso Kagaku Kogyo Co., Ltd. for complimentary supply of the 3YSZ powder used in Materials B and C in this study.

References

1. Ibrahim IA, Mohamed FA, Lavernia EJ (1991) Particulate reinforced metal matrix composites—a review. J Mater Sci 26:1137–1156
2. Lloyd DJ (1994) Particle reinforced aluminium and magnesium matrix composites. Int Mater Rev 39:1–23
3. Tjong SC (2007) Novel nanoparticle-reinforced metal matrix composites with enhanced mechanical properties. Adv Eng Mater 9:639–652
4. Borgonovo C, Apelian D (2011) Manufacture of aluminum nanocomposites: a critical review. Mater Sci Forum 678:1–22

5. Ma K, Lavernia EJ, Schoenung JM (2017) Particulate reinforced aluminum alloy matrix composites—a review on the effect of microconstituents. Rev Adv Mater Sci 48:91–104
6. Han BQ, Ye J, Tang F, Schoenung J, Lavernia EJ (2007) Processing and behavior of nanostructured metallic alloys and composites by cryomilling. J Mater Sci 42:1660–1672
7. Li Y, Zhang Z, Vogt R, Schoenung JM, Lavernia EJ (2011) Boundaries and interfaces in ultrafine grain composites. Acta Mater 59:7206–7218
8. Jiang L, Ma K, Yang H, Li M, Lavernia EJ, Schoenung JM (2014) The microstructural design of trimodal aluminum composites. JOM 66:898–908
9. Ye J, Schoenung JM (2004) Technical Cost modeling for the mechanical milling at cryogenic temperature (cryomilling). Adv Eng Mater 6:656–664
10. Ye J, He J, Schoenung JM (2006) Cryomilling for the fabrication of a particulate B_4C reinforced Al nanocomposite: Part I. Effects of process conditions on structure. Metall Mater Trans A 37:3099–3109
11. Ye J, Han BQ, Lee Z, Ahn B, Nutt SR, Schoenung JM (2005) A tri-modal aluminum based composite with super-high strength. Scripta Mater 53:481–486
12. Li Y, Zhao YH, Ortalan V, Liu W, Zhang ZH, Vogt RG, Browning ND, Lavernia EJ, Schoenung JM (2009) Investigation of aluminum-based nanocomposites with ultra-high strength. Mater Sci Eng A 527:305–316
13. Vogt RG, Zhang Z, Topping TD, Lavernia EJ, Schoenung JM (2009) Cryomilled aluminum alloy and boron carbide nano-composite plate. J Mater Process Technol 209:5046–5053
14. Yang H, Topping TD, Wehage K, Jiang L, Lavernia EJ, Schoenung JM (2014) Tensile behavior and strengthening mechanisms in a submicron B_4C-reinforced Al trimodal composite. Mater Sci Eng A 616:35–43
15. Zhang Z, Topping TD, Li Y, Vogt R, Zhou Y, Haines C, Paras J, Kapoor D, Schoenung JM, Lavernia EJ (2011) Mechanical behavior of ultrafine-grained Al composites reinforced with B_4C nanoparticles. Scripta Mater 65:652–655
16. Jiang L, Yang H, Yee JK, Mo X, Topping T, Lavernia EJ, Schoenung JM (2016) Toughening of aluminum matrix nanocomposites via spatial arrays of boron carbide spherical nanoparticles. Acta Mater 103:128–140
17. Ma ZY, Li YL, Liang Y, Zheng F, Bi J, Tjong SC (1996) Nanometric Si_3N_4 particulate-reinforced aluminum composite. Mater Sci Eng A 219:229–231
18. Zhang Z, Chen DL (2006) Consideration of Orowan strengthening effect in particulate-reinforced metal matrix nanocomposites: a model for predicting their yield strength. Scripta Mater 54:1321–1326
19. Garvie RC, Hannink RH, Pascoe RT (1975) Ceramic steel? Nature 258:703–704
20. Basu B (2005) Toughening of yttria-stabilised tetragonal zirconia ceramics. Int Mater Rev 50:239–256
21. Porter DL, Evans AG, Heuer AH (1979) Transformation-toughening in partially-stabilized zirconia (PSZ). Acta Metall 27:1649–1654
22. Hannink RHJ, Kelly PM, Muddle BC (2000) Transformation toughening in zirconia-containing ceramics. J Am Ceram Soc 83:461–487
23. Karihaloo BL (1991) Contribution of t→m Phase Transformation to the Toughening of ZTA. J Am Ceram Soc 74:1703–1706
24. Matsumoto Y, Hirota K, Yamaguchi O, Inamura S, Miyamoto H, Shiokawa N, Tsuji K (1993) Mechanical properties of hot isostatically pressed zirconia-toughened alumina ceramics prepared from coprecipitated powders. J Am Ceram Soc 76:2677–2680
25. De Aza AH, Chevalier J, Fantozzi G, Schehl M, Torrecillas R (2002) Crack growth resistance of alumina, zirconia and zirconia toughened alumina ceramics for joint prostheses. Biomaterials 23:937–945
26. Magnani G, Brillante A (2005) Effect of the composition and sintering process on mechanical properties and residual stresses in zirconia–alumina composites. J Eur Ceram Soc 25:3383–3392

27. Huang S, Binner J, Vaidhyanathan B, Brown P, Hampson C, Spacie C (2011) Development of nano zirconia toughened alumina for ceramic armor applications. In: Advances in ceramic armor VII. Wiley, pp 103–113
28. Geng L, Zheng ZZ, Yao CK, Mao JF, Imai T (2000) A new in situ composite fabricated by powder metallurgy with aluminum and nanocrystalline ZrO_2 particles. J Mater Sci Lett 19:985–987
29. Dutkiewicz J, Lityńska L, Maziarz W, Haberko K, Pyda W, Kanciruk A (2009) Structure and properties of nanocomposites prepared from ball milled 6061 aluminium alloy with ZrO_2 nanoparticles. Cryst Res Technol 44:1163–1169
30. Witkin DB, Lavernia EJ (2006) Synthesis and mechanical behavior of nanostructured materials via cryomilling. Prog Mater Sci 51:1–60
31. Hashemi-Sadraei L, Mousavi SE, Vogt R, Li Y, Zhang Z, Lavernia EJ, Schoenung JM (2012) Influence of nitrogen content on thermal stability and grain growth kinetics of cryomilled Al nanocomposites. Metall Mater Trans A 43:747–756
32. Zhao Y, Zhu Y, Lavernia EJ (2010) Strategies for improving tensile ductility of bulk nanostructured materials. Adv Eng Mater 12:769–778
33. Zhang Z, Dallek S, Vogt R, Li Y, Topping TD, Zhou Y, Schoenung JM, Lavernia EJ (2010) Degassing behavior of nanostructured Al and its composites. Metall Mater Trans A 41:532–541
34. Tang F, Hagiwara M, Schoenung JM (2005) Formation of coarse-grained inter-particle regions during hot isostatic pressing of nanocrystalline powder. Scripta Mater 53:619–624
35. Lucadamo G, Yang NYC, San Marchi C, Lavernia EJ (2006) Microstructure characterization in cryomilled Al 5083. Mater Sci Eng A 430:230–241
36. Topping TD, Ahn B, Li Y, Nutt SR, Lavernia EJ (2012) Influence of process parameters on the mechanical behavior of an ultrafine-grained Al alloy. Metall Mater Trans A 43:505–519
37. Lin Y, Wen H, Li Y, Wen B, Liu W, Lavernia EJ (2015) An analytical model for stress-induced grain growth in the presence of both second-phase particles and solute segregation at grain boundaries. Acta Mater 82:304–315
38. Topping TD (2012) Materials science and engineering. University of California, Davis, CA
39. Ye J (2006) Materials science and engineering. University of California, Davis
40. Yin Z, Pan Q, Zhang Y, Jiang F (2000) Effect of minor Sc and Zr on the microstructure and mechanical properties of Al–Mg based alloys. Mater Sci Eng A 280:151–155
41. Kendig KL, Miracle DB (2002) Strengthening mechanisms of an Al-Mg-Sc-Zr alloy. Acta Mater 50:4165–4175
42. Dobatkin S, Estrin Y, Zakharov V, Rostova T, Ukolova O, Chirkova A (2009) Improvement in the strength and ductility of Al-Mg-Mn alloys with Zr and Sc additions by equal channel angular pressing. Int J Mater Res 100:1697–1704
43. Lityñska L, Abou-Ras D, Kostorz G, Dutkiewicz J (2006) TEM and HREM study of Al_3Zr precipitates in an Al-Mg-Si-Zr alloy. J Microsc 223:182–184
44. Nes E (1972) Precipitation of the metastable cubic Al_3Zr-phase in subperitectic Al-Zr alloy. Acta Metall 20:499–506
45. Newbery AP, Ahn B, Pao P, Nutt S, Lavernia E (2007) A ductile UFG Al Alloy via cryomilling and quasi-isostatic forging. Adv Mater Res 29–30:21–29
46. Hall EO (1951) The deformation and ageing of mild steel: III discussion of results. Proc Phys Soc Sect B 64:747
47. Petch NJ (1953) The cleavage strength of polycrystals. J Iron Steel Inst 174:25–28
48. Hansen N (2004) Hall–Petch relation and boundary strengthening. Scripta Mater 51:801–806
49. Davis JR (1993) Aluminum and aluminum alloys. ASM International, Materials Park
50. Tsukuma K, Ueda K, Matsushita K, Shimada M (1985) High-temperature strength and fracture toughness of Y_2O_3-partially-stabilized ZrO_2/Al_2O_3 composites. J Am Ceram Soc 68: C-56–C-58

Fatigue Crack Growth Resistance of Titanium Metal Matrix Composites

H. Stanley, M. Dear, T. J. A. Doel and P. Bowen

Abstract Fatigue crack growth resistance has been assessed in continuous SCS-6 fibre reinforced Ti-6Al-4V composites from unbridged defects at room temperature, 300 and 450 °C in air and in vacuum. The initial applied stress intensity factor range (ΔK_{app}) (on the unbridged crack depth) was used to quantify the crack arrest/catastrophic failure (CA/CF) transitions for all test conditions. All tests were conducted at stress ratios of 0.1, 0.5 and 0.7 and at a frequency of 4 Hz. Fatigue crack growth tests revealed that increased temperature results in increased fatigue crack growth rates. The CA/CF transition reduced as the temperature increased from room temperature to 300 °C and to 450 °C. The CA/CF transition of the composite is greatly affected by the initial ΔK_{app} value due to its effect on fibre failures. Crack bifurcation appears to be a function of test piece geometry and not a function of test temperature or the initial ΔK_{app} value. Tests in vacuum demonstrate that fatigue crack growth rates are reduced. All cracks in vacuum grew in single mode I. Longer fibre pull out lengths were observed in tests performed in vacuum compared with air for similar conditions, especially during 0.5 mm of growth from the notch. These fibres are deduced to bridge the crack for a long time, suggesting that the fibres in the first few rows are important to reduce crack growth rates and to promote crack arrest.

Keywords Fatigue crack growth · Titanium metal matrix composites
Temperature · Environmental effects · Crack arrest · Fibre pull out

---

H. Stanley (✉) · M. Dear · T. J. A.Doel · P. Bowen
School of Metallurgy and Materials, The University of Birmingham,
Edgbaston, Birmingham B15 2TT, UK
e-mail: hxs260@bham.ac.uk

T. S. Srivatsan et al. (eds.), *Metal-Matrix Composites Innovations, Advances and Applications*, The Minerals, Metals & Materials Series,
https://doi.org/10.1007/978-3-319-72853-7_7

Introduction

The constant need for increasing performance demands by aerospace systems calls for the development of new materials. Titanium metal matrix composites (Ti MMCs) reinforced with silicon carbide (SiC) fibres are of interest to the aerospace industry because of their high specific mechanical properties, especially specific strength and stiffness. They have the ability to decrease the weight of several key axi-symmetric components, and may allow radical changes in design, and thereby improve the overall efficiency of the engine. Previous studies have investigated their potential use within an aero engine, to replace monolithic alloys, but cost and complexities of the composite limit its use. Unlike its monolithic counterpart, Ti MMCs exhibit no unique relationship between the crack growth rates, da/dN, and the nominal applied stress intensity factor, ΔK_{app} [1–3]. Studies have shown that Ti MMCs exhibit fatigue crack growth resistance through the use of bridging fibres behind the crack tip. The bridging fibres resist the crack opening forces and can reduce the value of the crack-tip stress intensity range, ΔK_{tip} by providing crack closure forces, and can reduce fatigue crack growth rates with an increase in crack length, due to a greater number of fibres bridging the crack [1, 4–6]. If fibres remain intact, the propagating crack can arrest (CA), which is defined as da/dN $\leq 10^{-8}$ mm/cycle [7]. If some bridging fibres fracture the da/dN can increase, sometimes resulting in catastrophic failure (CF), therefore a transition between the two behaviours exists. This is known as the CA/CF transition and can be defined in terms of ΔK. This transition however is dependent on the number of bridging fibres in the crack wake as well as secondary factors such as the test temperature, environment, ΔK, and the stress ratio, R.

One possible area where Ti MMCs are being considered is in the compressor region of the engine where temperatures do not exceed 400 °C. One component that will benefit from the implementation of Ti MMCs is a bladed ring. By removing the need for the central bore the component will reduce its weight. This component will experience high stresses and cyclic loading and therefore Ti MMCs resistance to fatigue is of utmost importance. The circumferential fibres are able to support the hoop stress alone [7]. This paper attempts to characterise and discuss the fatigue crack growth resistance of a Ti-6Al-4V/SCS-6 composite at ambient and elevated temperatures, with particular attention to determining the CA/CF transitions. The effect of temperature, environment, and initial ΔK_{app} will be discussed, with some reference to pull out lengths and stress ratio.

Method

The Ti-6Al-4V/SCS-6 composite was of an 8-ply unidirectional layup with a fibre volume fraction of 35%. The precursor material was SCS-6 SiC fibre of 142 μm nominal diameter, incorporating a 3 μm carbon outer layer, with a Ti-6Al-4V physical vapour deposition coating. The metal coated fibre was laid up as a panel in a hexagonal array, with a ~150 μm cladding layer on either side. This was hot isostatic pressed to form the dense composite. Whilst the volume of the composite core was 35%, addition of the cladding layer reduced the overall volume fraction to ~27%.

Test pieces of 75 × 4 × 1.9 mm were cut for three point bend tests. A notch of 0.4 mm depth was cut into the middle of each test piece, perpendicular to the fibre direction using a 150 μm thick diamond blade, resulting in an initial crack depth to width ratio, a_0/W of 0.1.

The test pieces were loaded in three-point bend using a 60 mm span, giving a span to width ratio of 15:1. All tests were conducted using a servo-hydraulic machine operating in load control mode, using a 5kN range. The machine could be configured for testing at temperature and in air or vacuum as required. Load was applied using a sinusoidal waveform with maximum and minimum loads being at a constant level throughout each test. Tests were conducted at room temperature (20 °C), 300 and 450 °C, temperatures being monitored throughout using thermocouples attached to the test pieces.

Fatigue crack growth was monitored using the direct current potential difference (DCPD) technique. A constant current was applied to the test piece, and the resulting voltage across the notch was measured via spot welded platinum wires. As the crack length increased, the resistance increased, therefore according to the Ohm's Law, the voltage increased. A starting voltage of 1 mV was used across the notch and the V_0 value was adjusted in order to match the measured notch length with the calculated a_0. The recorded voltage was used to calculate the crack length using the equation [8].

$$\frac{a}{W} = -0.45 + 0.954\left(\frac{V}{V_0}\right) - 0.292\left(\frac{V}{V_0}\right)^2 + 0.038\left(\frac{V}{V_0}\right)^3$$

Stress ratio, R, values of 0.1, 0.5 and 0.7 and a frequency of 4 Hz was used. The tests performed at room temperature had initial ΔK_{app} values of between 18 and 30 MPa$\sqrt{m}$. Initial ΔK_{app} values between 14 and 23 MPa$\sqrt{m}$ were used for the test temperature of 300 °C. For tests conducted in air at 450 °C, initial ΔK_{app} values between 8 and 18 MPa$\sqrt{m}$ were used. Initial ΔK_{app} values between 15 and 30 MPa$\sqrt{m}$ were used to identify the effects that elevated temperature and environment has on fatigue crack growth resistance.

After each test the final crack length was measured from the fracture surface and checked against that given by the DCPD technique. The fracture faces were initially examined using low magnification optical images on a Leica stereo microscope, and they were then further examined using a Hitachi S4000 field emission gun scanning electron microscope (SEM) operating at 20 kV. All images were taken using the secondary electron detector. While examining the fracture faces, fibre pull out lengths were measured in the region of the mode I fatigue crack.

Results and Discussion

Fatigue Crack Growth Resistance

Due to the fibre bridging mechanism, it is not appropriate to present fatigue crack growth data as conventional da/dN versus nominal ΔK plots. The approach here is to compare the results in terms of da/dN versus crack length, a, for given values of the initial ΔK_{app}.

Note: The da/dN data in all figures and tables are stated for mode I crack lengths only. Bifurcation, where the crack can split and grow along the fibre/matrix interface, is observed in some tests within this paper. Bifurcation invalidates the DCPD calibration curve and hence data is shown only to the bifurcation length.

Effects of ΔK at a Single Test Temperature

Figures 1, 3 and 4 display the fatigue crack growth curves for room temperature, 300 and 450 °C. The CA/CF transitions are shown in Table 1 for each temperature. From Fig. 1, at room temperature, the test at an initial ΔK_{app} value of 18 MPa$\sqrt{m}$ reached crack arrest, where the effective ΔK reduced below matrix threshold levels. The test at an initial ΔK_{app} of 30 MPa$\sqrt{m}$ experienced stable crack growth rates until rapid increases in da/dN were observed prior to catastrophic failure. The two tests performed at initial ΔK_{app} values of 21 and 23 MPa$\sqrt{m}$ experienced very similar crack growth rates to the test that reached crack arrest but the fatigue crack

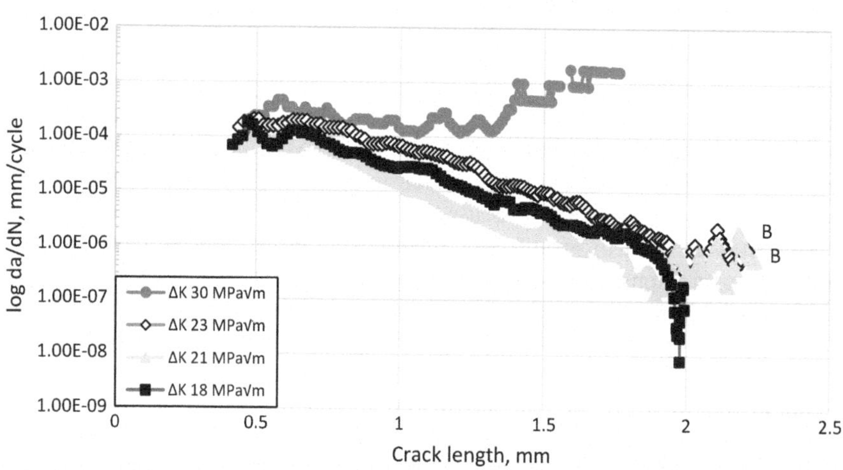

Fig. 1 Fatigue crack growth rate da/dN versus crack length for four different values of initial ΔK_{app} at $R = 0.1$ cycled at 4 Hz until failure or arrest at room temperature. *Note* The da/dN data for tests that bifurcated is only displayed up to the bifurcation crack length

Table 1 The initial ΔK_{app} values required for the CA/CF transition for each test temperature

Initial ΔK_{app} (MPa$\sqrt{m}$)	Temperature (°C)	Outcome
18	20	Arrest
21	20	Failure
14	300	Arrest
16	300	Failure
10	450	Arrest
12	450	Failure

in both of these eventually bifurcate at crack lengths of approximately 2.2 mm [9]. Therefore, the CA/CF transition at room temperature is estimated between 18 and 21 MPa$\sqrt{m}$. An image of bifurcation that occurred on the test at an initial ΔK_{app} of 21 MPa$\sqrt{m}$ is shown in Fig. 2.

Figure 3 displays the results from the fatigue crack growth tests at 300 °C. The tests with at initial ΔK_{app} value of 18 MPa$\sqrt{m}$ and below experienced periods of decreasing crack growth rates which almost led to crack arrest. Except for the test where crack arrest was actually achieved at an initial ΔK value of 14 MPa$\sqrt{m}$, the fatigue cracks bifurcated in each case. The test at an initial ΔK_{app} of 23 MPa$\sqrt{m}$ experienced secondary cracking, where a second crack formed but it continued to grow in mode I direction until it bifurcated at a crack length of 2.25 mm.

The tests at 450 °C exhibited a similar trend to those at 300 °C, shown in Fig. 4. Tests performed at initial ΔK_{app} values of 8 and 10 MPa$\sqrt{m}$ are deemed to have reached crack arrest, while for those tested at values greater than 12 MPa$\sqrt{m}$ the fatigue crack bifurcated and the test piece failed. Therefore, the CA/CF transition is estimated between 10 and 12 MPa$\sqrt{m}$. The three tests where the crack bifurcated and failure occurred saw fatigue crack growth rates decrease after 0.55 mm of crack growth. At a ΔK_{app} of 18 MPa$\sqrt{m}$, da/dN continued to decrease at a stable rate until a crack length of 1.6 mm was reached, where the fatigue crack growth rates increased until final failure. For the tests performed at initial ΔK_{app} values of 12 and 16 MPa$\sqrt{m}$, da/dN was much less stable and a rapid increase in da/dN was observed near the failure crack length.

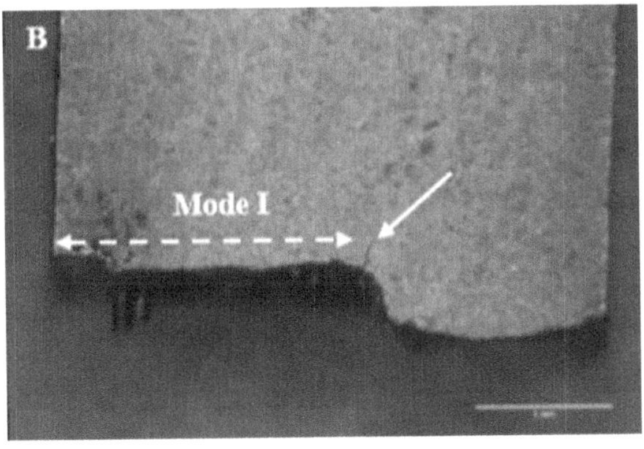

Fig. 2 Fractography of the test piece at an initial ΔK_{app} of 21 MPa$\sqrt{m}$. The dotted line indicates the mode I crack growth and the arrow points to the point of bifurcation [9]

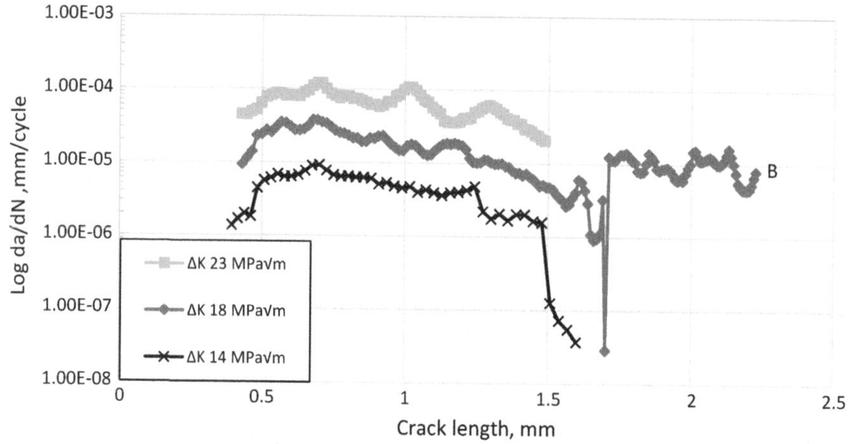

Fig. 3 Fatigue crack growth rate da/dN versus crack length for four different values of initial ΔK_{app} at R = 0.1 cycled at 4 Hz until failure or arrest at 300 °C. *Note* The da/dN data for tests that bifurcated is only displayed up to the bifurcation crack length. Data for initial ΔK_{app} 23 MPa$\sqrt{m}$ data is only displayed up to the length where a secondary crack formed

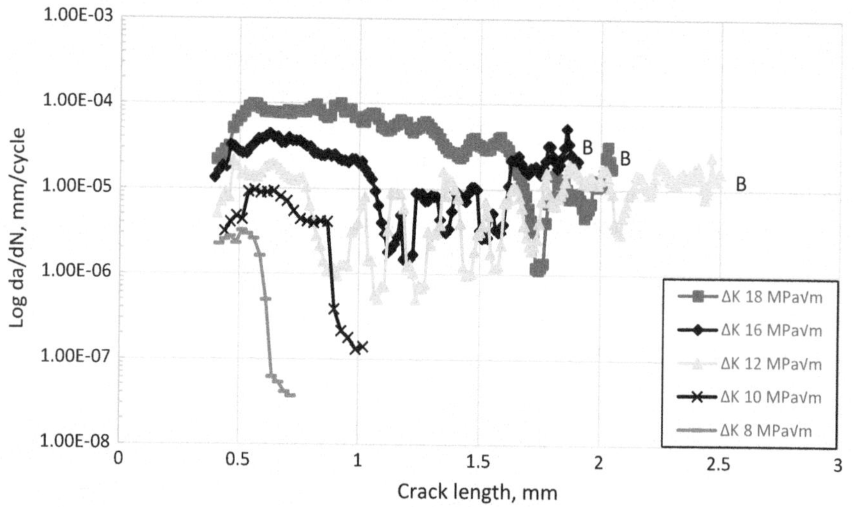

Fig. 4 Fatigue crack growth rate da/dN versus crack length for five different values of initial ΔK_{app} at R = 0.1 cycled at 4 Hz until failure or arrest at 450 °C. *Note* The da/dN data for tests that bifurcated is only displayed up to the bifurcation crack length

There is an initial increase in da/dN with increasing crack length. This occurs because the fatigue crack needs to grow approximately 0.25 mm from the notch to grow past one row of fibres and experience the effects of bridging fibres. After this initial increase the da/dN stabilises and starts to reduce. This reduction is an effect

Table 2 Number of cycles to arrest/failure for a number of tests across the three temperatures

Condition	Crack length (mm)	Number of cycles to arrest/failure
20 °C/ΔK = 23 MPa√m	2.21 (B)	1,285,448 579,288 (B)
20 °C/ΔK = 18 MPa√m	1.99	1,352,376
300 °C/ΔK = 18 MPa√m	2.22 (B)	260,265 214,296 (B)
300 °C/ΔK = 14 MPa√m	1.6	929,640
450 °C/ΔK = 16 MPa√m	1.9 (B)	203,400 186,660 (B)
450 °C/ΔK = 10 MPa√m	1.02	644,000

of the crack growing past fibres which are left bridging the crack wake. All fatigue crack growth results at room temperature, 300 and 450 °C indicate that a decrease in da/dN rates is observed with increasing crack length. For any given temperature, decreasing the initial ΔK_{app} increases the specimen lifetime (Table 2). Crack arrest is observed at the lowest values of initial ΔK_{app} used in this paper. The main mechanism promoting fatigue crack growth resistance is the presence of intact bridging fibres in the crack wake which shield the crack tip from the full effect of the ΔK_{app} [1]. The number and effect of these intact fibres in the crack wake is controlled by several factors; the fibre strength and distribution, the number of bridging fibres, the fibre matrix interfacial shear strength [10], the initial ΔK_{app}, manufacturing method and test temperature [3]. The sharp decreases in da/dN observed by all tests which arrest can be attributed to ΔK_{app} at the crack tip (ΔK_{tip}) falling below the threshold stress intensity range for the matrix, (ΔK_{th}). Bifurcation occurred in test pieces at each temperature at very similar crack lengths. This suggests bifurcation is not a function of test temperature or initial ΔK_{app}, but is a function of test piece geometry.

Effect of Temperature

Figure 5 shows the effect of test temperature on the fatigue crack growth rates at R = 0.1 in air and at an initial ΔK_{app} value of 18 MPa√m. The comparison of the curves suggest that the initial crack growth rates at all test temperatures are very similar. At a test temperature of 20 °C crack growth rates (da/dN) decreased until the crack arrested at a crack length of 1.99 mm. Increasing the test temperature to 300 and 450 °C increased the growth rates after 1.4 mm of crack growth until catastrophic failure was observed. Cracks in tests at 300 and 450 °C grew in mode I up to 2.22 and 2.04 mm respectively before bifurcating. The fatigue curves suggest that there is not much difference between test temperatures of 20 and 300 °C as the latter almost arrested at a crack length of 1.7 mm. This is confirmed by the similar crack growth rates at room temperature and 300 °C up to a crack length of 1.5 mm.

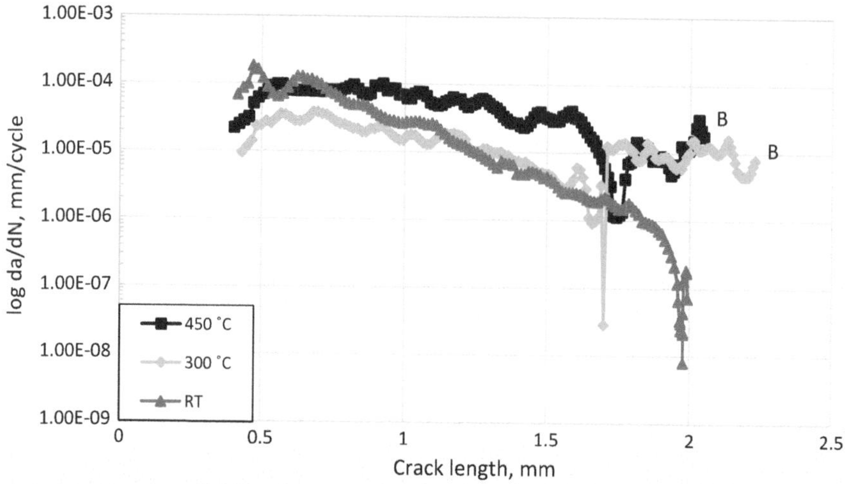

Fig. 5 Fatigue crack growth rate da/dN versus crack length for three different temperatures: 20, 300 and 450 °C. All tested at an initial ΔK_{app} of 18 MPa$\sqrt{m}$ at R = 0.1 and 4 Hz. *Note* The da/dN data for tests that bifurcated is only displayed up to the bifurcation crack length

As the test temperature increased, the number of cycles to failure decreased and the final crack length increased (Table 3). Similar behaviour is also observed at R = 0.5 and 0.7 (Table 3), however the number of cycles to failure increased as the R ratio increased. It is difficult to find a CA/CF transition at higher R ratios at 450 °C as smaller initial ΔK_{app} values failed to initiate from the notch [11].

Table 1 shows that increasing the test temperature decreases the initial ΔK_{app} value at which the CA/CF transition occurs and increases the likelihood of bifurcation. There are multiple reasons why the fatigue crack growth resistance is greater at lower temperatures. Firstly, at elevated temperatures the matrix material itself may experience reduced fatigue crack growth resistance, increasing the fatigue crack growth rates of the composite [2]. Secondly, changes to the interfacial properties between the matrix and fibre during testing could arise from changes in the residual stresses or from effects of oxidation at the interface. The compressive stress exerted onto the fibres by the matrix is reduced at 450 °C due to the thermal expansion coefficient mismatch and therefore the likelihood of the fatigue crack propagating at the fibre/matrix interface is increased. If the interfacial strength decreases at higher temperatures, provided the fibre fracture strength stays the same, models based on ceramic matrix composites would suggest pull out lengths increase [12]. There is some indication of this from individual fibre tests and their interfacial properties, however as they are measured at room temperature and not at the operating temperatures, their results must be analysed with caution and an effective method of evaluating interfacial strengths at higher temperatures is needed. Pull out lengths for tests that reached crack arrest for each corresponding temperature are displayed in Table 4. There appears to be very little difference at all temperatures between average pull out lengths.

Table 3 Comparison of crack lengths and number of cycles for tests at different temperatures, environment conditions, R ratios and initial ΔK_{app} at a frequency of 4 Hz

Temperature (°C)	Environment	R Ratio	Initial ΔK_{app}									
			6	8	10	12	14	16	18	21	22	24
Crack length (mm)												
20	Air	0.1	–	–	–	–	–	–	1.99	2.24 (B)	–	–
450	Air	0.1	–	–	1.02	2.5 (B)	–	–	–	–	–	–
450	Vacuum	0.1	–	–	–	–	–	–	–	–	2.32	1.96
20	Air	0.5	–	–	–	–	1.37	2.22	–	–	–	–
450	Air	0.5	–	2.41[a]	–	–	–	–	–	–	–	–
20	Air	0.7	–	–	–	1.64	1.83	–	–	–	–	–
450	Air	0.7	2.36[a]	–	–	–	–	–	–	–	–	–
Number of cycles (×100)												
20	Air	0.1	–	–	–	–	–	–	1352	1602 (B)	–	–
450	Air	01	–	–	644	634 (B)	–	–	–	–	–	–
450	Vacuum	0.1	–	–	–	–	–	–	–	–	1636	1000
20	Air	0.5	–	–	–	–	399	1708	–	–	–	–
450	Air	0.5	–	915[a]	–	–	–	–	–	–	–	–
20	Air	0.7	–	–	–	810	1277	–	–	–	–	–
450	Air	0.7	1575[a]	–	–	–	–	–	–	–	–	–

The two tests representing crack arrest and catastrophic failure have been displayed

[a]Crack arrest did not occur and therefore the lowest initial ΔK value has been used

Note The data for tests that bifurcated is only displayed up to the bifurcation crack length and is denoted by ([a])

Table 4 Showing the average fibre pull out length and maximum pull out length measured from one side of the fracture face on a SEM for three test temperatures

Condition	Average pull out (mm)	S.D (mm)	Max pull out (mm)
20 °C/$\Delta K = 18$ MPa$\sqrt{m}$	0.076	0.029	0.14
300 °C/$\Delta K = 14$ MPa$\sqrt{m}$	0.095	0.069	0.31
450 °C/$\Delta K = 10$ MPa$\sqrt{m}$	0.078	0.052	0.19

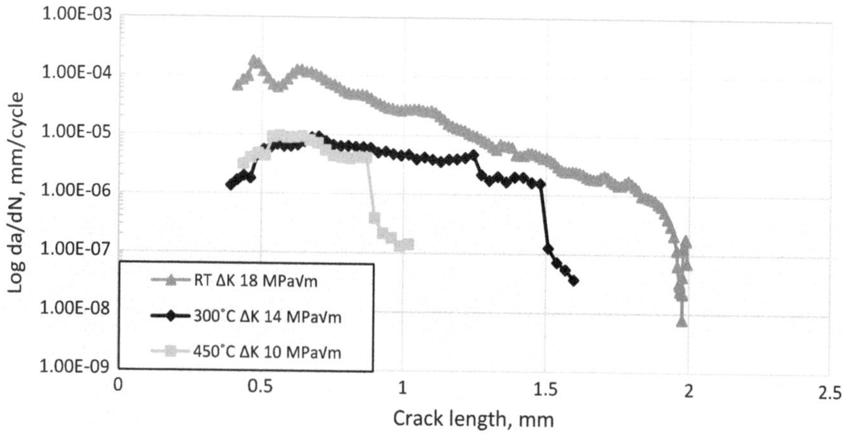

Fig. 6 Fatigue crack growth rate da/dN versus crack length for three different temperatures: 20, 300 and 450 °C and their corresponding crack arrest values of initial ΔK_{app}. R = 0.1 and 4 Hz

Figure 6 displays tests from the three temperatures and their corresponding crack arrest fatigue curves. Note the data displayed is data from the highest initial ΔK_{app} that arrested. It is clear that the maximum initial ΔK_{app} used to achieve crack arrest decreases progressively as the temperature is increased from room temperature to 300 and 450 °C.

Table 4 displays the average and maximum pull out lengths measured from the tests that arrested at the three temperatures. Despite the reduced residual stresses there appears to be no difference between room temperature and 450 °C. Figure 7 highlights the similarities of reduced pull out lengths at room temperature and 450 °C in air.

Effect of Environment (Air vs. Vacuum) at 450 °C

Testing in vacuum has a clear effect on fatigue crack growth rates when compared with tests in air and at room temperature (Fig. 8). The rates are much slower, by about two orders of magnitude in vacuum than tests performed in air at an R = 0.1 and similar initial ΔK_{app} values. Apart from the initial 0.1 mm of crack growth the

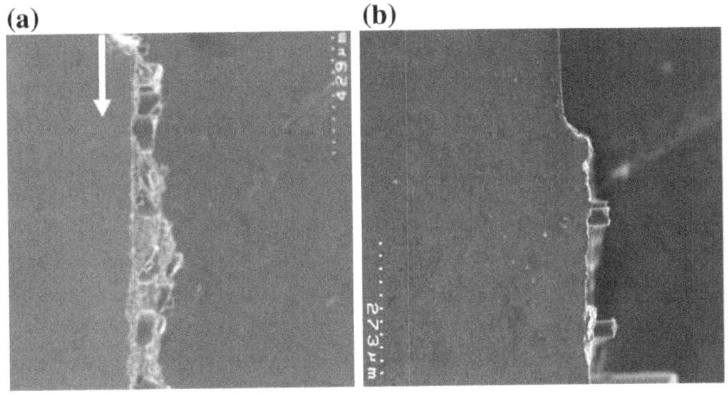

Fig. 7 SEM images showing the pull out lengths of the two tests performed at **a** Room temperature and an initial ΔK_{app} value of 18 $MPa\sqrt{m}$, and **b** 450 °C at an initial ΔK_{app} value of 10 $MPa\sqrt{m}$

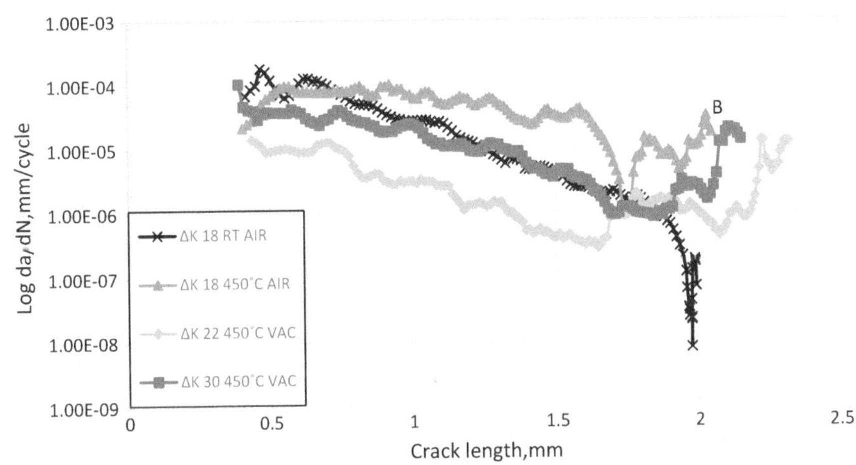

Fig. 8 Fatigue crack growth rate da/dN versus crack length for different values of initial ΔK_{app} at R = 0.1 cycled at 4 Hz until failure or arrest at room temperature and 450 °C in air and vacuum. *Note* The da/dN data for tests that bifurcated is only displayed up to the bifurcation crack length

test with an initial ΔK_{app} of 18 $MPa\sqrt{m}$ in air at 450 °C has higher crack growth rates than the test with an initial ΔK_{app} value of 30 $MPa\sqrt{m}$ in vacuum for the entirety of the test.

Although as of yet a CA/CF transition has not been established in vacuum at 450 °C, the behaviour of the lowest initial ΔK_{app} value of 22 $MPa\sqrt{m}$ applied to date suggests that crack arrest will occur at a much higher initial ΔK_{app} value than air. It is obvious from the fatigue crack growth curves that for testing in vacuum decreases fatigue crack growth rates compared to air.

When comparing tests performed in vacuum at 450 °C and tests performed in air at room temperature the test with a much higher initial ΔK_{app} value of 30 MPa$\sqrt{m}$ in vacuum has slightly reduced crack growth rates compared to that tested at room temperature which grows to crack arrest. This is the case until the cracks reach a length of 1.9 mm when fibre failures lead to fracture of the test piece in vacuum. Brisset [2] also found da/dN similar in vacuum at this temperature and in air at room temperature, finding the CA/CF for both to be the same.

Fracture surfaces and crack growth profiles display differences between tests in air and vacuum. It is evident from Fig. 9 that the tests performed in vacuum have on average much longer pull out lengths than tests performed in air (Fig. 7, Table 5). One of the main differences between the two are the position of these bridging fibres. It is clear from the images that there is limited pull out in the first few rows of fibres for the test performed in air, compared with the long pull out apparent in vacuum. These fibres will have been bridging the crack for a long time in vacuum, and therefore reducing the growth rates shown. These experiments need to be repeated with acoustic emission to identify more precisely when fibre failures occur. To date, deductions from sharp increases in da/dN are used to identify fibre failures.

Figure 10 displays fatigue crack growth data for all tests in vacuum at 450 °C. The initial ΔK_{app} ranged from 22 to 30 MPa$\sqrt{m}$. All tests initially have crack

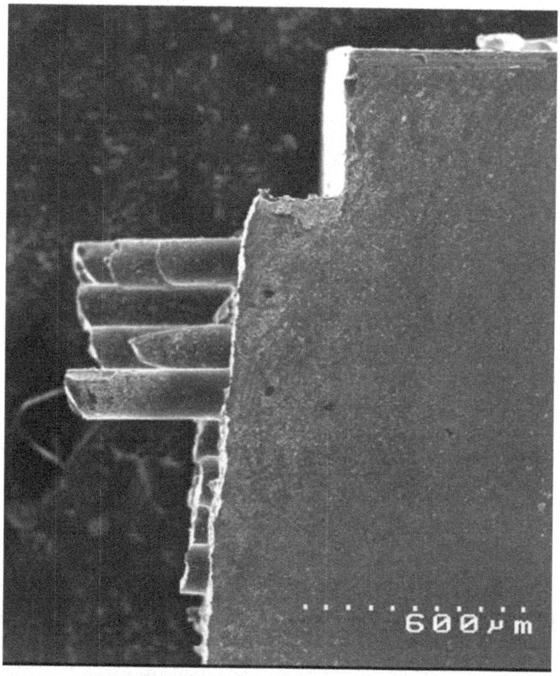

Fig. 9 SEM images showing the pull out lengths of the test performed at 450 °C and an initial ΔK_{app} value of 30 MPa$\sqrt{m}$

Table 5 Showing the average fibre pull out length and maximum pull out length measured from one side of the fracture face on a SEM for three tests

Condition	Average pull out (mm)	S.D (mm)	Max pull out (mm)
20 °C/AIR ΔK = 18 MPa√m	0.076	0.029	0.14
450 °C/AIR ΔK = 10 MPa√m	0.078	0.052	0.19
450 °C/VAC ΔK = 22 MPa√m	0.26	0.16	0.46

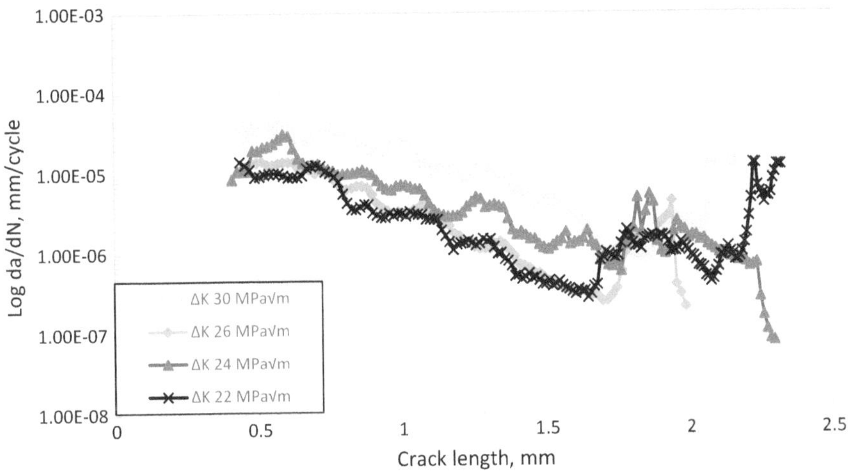

Fig. 10 Fatigue crack growth rate da/dN versus crack length for four different values of initial ΔK_{app} at R = 0.1 cycled at 4 Hz until failure or arrest at 450 °C in vacuum

growth rates scaled to the initial ΔK_{app}. All tests had very similar starting crack growth rates, but as the initial ΔK_{app} decreased, so too did the rate of growth. Even at the highest initial ΔK_{app}, crack growth rates decreased until final failure. All tests grew to a crack length of ~1.6 mm with decreasing crack growth rates. Figure 11 shows the effect of testing in vacuum, even at a higher test temperature, discussed earlier, and still has reduced da/dN compared to air. Earlier, the test temperature of 450 °C was found to have a reduced effect on the fatigue crack growth resistance of the composite. Taking this into account, Fig. 11 displays the benefits of vacuum. For the same initial ΔK_{app}, those tested in vacuum have higher resistance to fatigue cracks than those in air. This is due to the bridging fibres in the first few rows, which seem to bridge the crack for a very long time in vacuum.

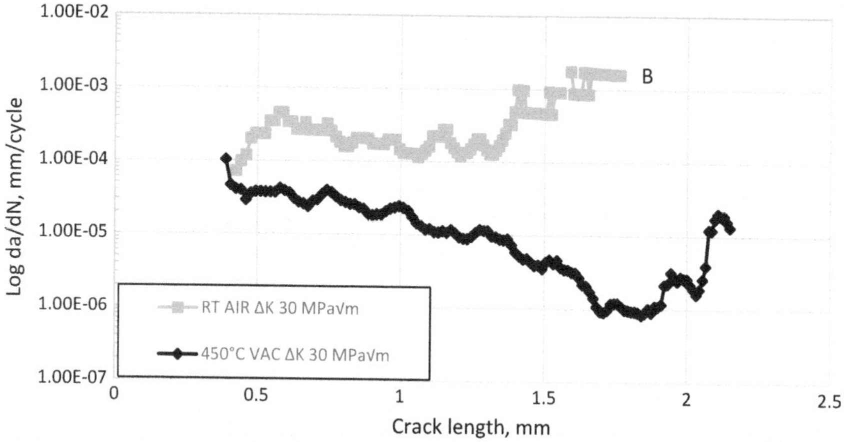

Fig. 11 Fatigue crack growth rate da/dN versus crack length for an initial ΔK_{app} 30 MPa$\sqrt{m}$ at room temperature in air and an ΔK_{app} 30 MPa$\sqrt{m}$ at 450 °C in vacuum. *Note* The da/dN data for test that bifurcated is only displayed up to the bifurcation crack length

Effect of Stress Ratio, R, at Room Temperature and 450 °C

Table 3 displays results from fatigue tests at stress ratios 0.5 and 0.7, both at room temperature and 450 °C. At both temperatures, an increase in stress ratio decreases the stress intensity factor range at which crack arrest can occur. A CA/CF transition was not found at both R ratio values at 450 °C as at initial ΔK_{app} values of 4 and 6 MPa$\sqrt{m}$ for 0.5 and 0.7 respectively, a fatigue crack did not initiate from the notch. At room temperature, the increase in stress ratio from 0.1 to 0.7 caused a decrease in the initial ΔK_{app} at which fatigue crack arrests. This trend agrees with the previous work by Brett et al. [5] and Cardona et al. [4].

The maximum applied stress intensity factor, K_{max}, after the crack has grown past the first row of fibres can also be used to quantify the CA/CF transition. This value, which is the same for all stress ratios at a given test temperature, should not be exceeded in order to keep the fibres intact and therefore to bridge the crack. Koval [13] found that at room temperature as the stress ratio increased, the K_{max} required for CA to occur increased. This suggests that the CA/CF transition is not only determined by K_{max}, but also by the initial ΔK_{app}.

Figure 12 displays fatigue crack growth data from tests performed at the same initial ΔK_{app} value (10 MPa$\sqrt{m}$) for three R ratios (0.1, 0.5 and 0.7) [11]. Higher stress ratios decreases the resistance of the composite to fatigue crack growth. The only test to arrest was the one with a stress ratio of 0.1. As stated earlier, sharp excursions in da/dN are indicative of fibre failures. The two higher stress ratios promote increased fibre fracture.

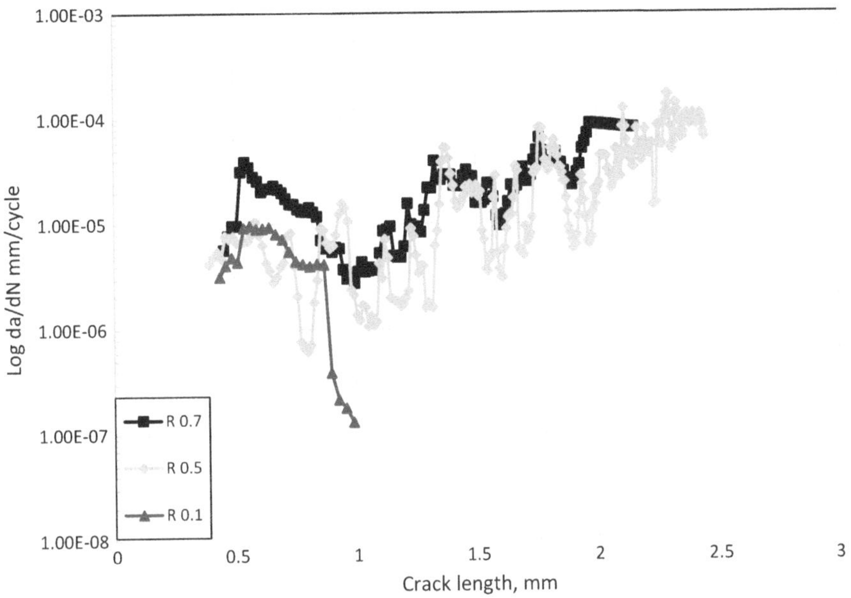

Fig. 12 Fatigue crack growth rate da/dN versus crack length for three different values of stress ratio, R, for tests in air at 450 °C. All tested at an initial ΔK_{app} of 10 MPa$\sqrt{m}$ and 4 Hz

Conclusions

The CA/CF transition is governed by the number of intact bridging fibre in the crack wake. A crack is deemed to have reached crack arrest if the growth rates fall below da/dN $\leq 10^{-8}$ mm/cycle. From the crack growth resistance curves it can be deduced that in tests which arrest few, if any, fibres have failed. Above this transition many fibres fail leading to the eventual fracture of the test piece.

At a given test temperature, initial ΔK_{app} determines the life of the composite. The higher the ΔK_{app} the more likely the test will fail. Bifurcation however, seems to be a function of test piece geometry and not affected by the initial loading conditions or test temperature.

There are modest effects of fatigue crack growth resistance between room temperature and 300 °C. Crack growth rates are very similar at a given crack length, up to 1.6 mm. Increasing the test temperature to 450 °C displays a much larger difference compared with room temperature. The CA/CF transition is greatly reduced at 450 °C, indicating temperature causes a change to the matrix resistance to fatigue cracks.

Testing in vacuum is advantageous compared with testing in air. Although a CA/CF transition is yet to be found, the fatigue crack growth data suggest this will be higher than in air at the same temperature. Individual fibre push out tests are required to assess the interfacial properties after tests in vacuum.

Fibre pull out lengths seem to be unaffected by an increase in temperature when observing test pieces that have arrested. The stark difference is observed in those tested in vacuum, where long pull out is observed in the first few rows of fibres, suggesting these rows are important in reducing crack growth rates.

An increase in stress ratio decreases the initial stress intensity factor range at which crack arrest can occur.

References

1. Cotterill P, Bowen P (1993) Fatigue crack growth in a fibre-reinforced titanium MMC at ambient and elevated temperatures. Compos 24(3):214–221
2. Brisset F, Bowen P (1998) Fatigue crack grovvth in fibre reinforced titanium MMC laminate at room and elevated temperatures. Mater Sci Technol 14(7):651–657
3. Bowen P (1996) Characterization of crack growth resistance under cyclic loading in the presence of an unbridged defect in fiber-reinforced titanium metal matrix composites. In: Life prediction methodology for titanium matrix composites. ASTM International
4. Cardona DC, Barney C, Bowen P (1996) Modeling and prediction of crack arrest in fiber-reinforced composites. In: Life prediction methodology for titanium matrix composites. ASTM International
5. Brett R, Cotterill P, Bowen P (1996) The influence of loading levels, temperature and environment on fibre failure during the fatigue of an SiC fibre reinforced Ti-MMC. Int J Fatigue 18(1):1–8
6. Doel T, Cardona D, Bowen P (1998) Fatigue crack growth in selectively reinforced titanium metal matrix composites. Int J Fatigue 20(1):35–50
7. Barney C, Cardona D, Bowen P (1998) Fatigue crack growth resistance from unbridged defects in continuous fibre reinforced metal matrix composites. Int J Fatigue 20(4):279–289
8. Ibbotson AR (1994) Fatigue crack growth in continuous fibre reinforced titanium alloy matrix composites. University of Birmingham
9. Dear MN (2016) Fatigue in SiC fibre reinforced titanium metal matrix composites. University of Birmingham
10. Zamperini S (2002) Effects of cyclic loading on fibre reinforced titanium metal matrix composites. University of Birmingham
11. Grantham-Wright H (2017) Fatigue crack growth resistance in fibre-reinforced titanium metal matrix composites. University of Birmingham
12. Liu J, Bowen P (1999) Bridging mechanisms in fibre-reinforced Ti MMCs. ROLLS ROYCE PLC-REPORT-PNR
13. Koval O (2016) Fatigue in SiC fibre reinforced titanium metal matrix composites. University of Birmingham

Experimental Optimization of Dry Sliding Wear Behavior of Titanium Matrix Composites Using Taguchi Methods

Koutarou Hattori, Shogen Hirami, Yoshiko Hasegawa, Hiroshi Izui and Yoshiki Komiya

Abstract Titanium and its alloys have high specific tensile strength and exhibit poor wear resistance. To improve their wear resistance, ceramic-particulate reinforced titanium matrix composites (TMCs) are fabricated by a spark plasma sintering process (SPS). The wear behavior of the TMCs depends on various factors, such as the matrix and reinforcement materials, reinforcement volume fraction, applied load, sliding load, sliding distance, and sliding velocity. The influencing factors on the wear behavior of TMCs in Taguchi methods are classified as either noise factors or control factors. Noise factors include load, sliding velocity, and sliding distance, and control factors include type, shape, particle size of matrix material, type of reinforcement material, and reinforcement volume fraction. In this study, the control factors were selected for optimization by Taguchi methods. Experiments were conducted systematically based on an L_{18} orthogonal array of the methods. Wear tests were carried out using a three-ball-on-disk machine. The results indicated that the wear behavior of TMCs was affected by three factors, namely, the type of reinforcement material, the reinforcement volume fraction, and the type of matrix material, but was not affected by two factors, namely, the shape of the matrix material and the particle size of the matrix material. An Si_3N_4-reinforced hydride-dehydrate Ti matrix composite with 10 vol.% reinforcement showed good wear resistance.

K. Hattori (✉) · S. Hirami · H. Izui · Y. Komiya
Nihon University, Tokyo, Japan
e-mail: csko17025@g.nihon-u.ac.jp

S. Hirami
e-mail: cssi13107@g.nihon-u.ac.jp

H. Izui
e-mail: izuihi@aero.cst.nihon-u.ac.jp

Y. Komiya
e-mail: komiya.y@aero.cst.nihon-u.ac.jp

Y. Hasegawa
Hasegawa Professional Engineer Office, Yokohama, Japan
e-mail: yohase03@nifty.com

T. S. Srivatsan et al. (eds.), *Metal-Matrix Composites Innovations, Advances and Applications*, The Minerals, Metals & Materials Series,
https://doi.org/10.1007/978-3-319-72853-7_8

Keywords Titanium matrix composites · Spark plasma sintering
Wear behavior · Taguchi methods

Introduction

In recent years, the application of lightweight carbon fiber reinforced plastic (CFRP) to body and engine parts of aircraft has been increasing in order to reduce fuel consumption. Titanium and its alloys have excellent high specific strength and corrosion resistance, and the demand for these materials is increasing because of their compatibility with CFRP [1, 2]. However, titanium and its alloys are inferior in terms of processing costs and wear resistance. One way to overcome such problems is to employ powder metallurgy. By using powder metallurgy, it becomes possible to mold the material into a shape close to the desired finished shape (near net shape), thus reducing the processing cost. In addition, it is generally considered that wear resistance is improved by making composites, and it is possible to prepare a composite material in which hard ceramic-particulates are added.

The wear behaviour is influenced by various factors, such as load, sliding speed, sliding distance, lubrication, atmosphere, and material properties. Using Taguchi methods, the influencing factors on the wear resistance of the titanium matrix composites examined in this study are classified into noise factors and control factors. The noise factors include load, sliding speed, sliding distance, the presence or absence of lubrication, and the temperature and humidity of the atmosphere. On the other hand, the control factors include the type of reinforcement material and matrix material, reinforcement volume fraction, and the shape and particle size of the powder. By using Taguchi methods, it is possible to greatly reduce the number of experiments, to clarify the influence of each factor, and to optimize each factor. In previous work on material evaluation using Taguchi methods, there have been many studies focusing on noise factors. Although these studies looked at various materials, the amount of wear mainly tends to increase with increasing load and sliding distance [3, 4]. Generally, it is considered that the amount of wear increases with increasing load and sliding distance [5]. Therefore, in this study, we only focused on the control factors that affect the wear resistance of titanium matrix composites. Previous studies have shown that wear resistance was improved by adding hard ceramic-particulate TiC and TiB_2 to pure titanium [6]. In this study, in addition to pure titanium, Ti-6Al-4V titanium alloy was used as the matrix, and hard ceramic-particulates of B_4C, SiC, Si_3N_4 were newly used in addition to TiC and TiB serving as the reinforcement. We used pure titanium and Ti-6Al-4V titanium alloy as the matrix, each in the form of spherical gas-atomized powder (GA) and angular hydride-dehydride powder (HDH). In both cases, two particle sizes were used (<45 μm, <150 μm). In this study, the control factors were allocated to an L_{18} orthogonal array in the Taguchi methods, with the aim of clarifying their influencing factors and improving the wear resistance of the titanium matrix composites.

Experimental Procedures

Fabrication of TMCs

We used four types of pure titanium powders and four types of Ti-6Al-4V titanium alloy powders as the matrixes: GA45 (nominal particle size <45 μm, from Osaka Titanium Technologies Co., Ltd., Japan) and GA150 (nominal particle size <150 μm, from Osaka Titanium Technologies Co., Ltd., Japan) prepared by the gas-atomized process, and HDH45 (nominal particle size <45 μm from Toho Titanium Co., Ltd., Japan) and HDH150 (nominal particle size <150 μm from Toho Titanium Co., Ltd., Japan) prepared by the hydride-dehydride process. We used six types of reinforcements: TiC (particle size 1.80 μm, Japan New Metals Co., Ltd., Japan), TiB_2 (particle size 1.81 μm, Japan New Metals Co., Ltd., Japan), B_4C (particle size 0.5 μm, Kojundo Chemical Laboratory Co., Ltd., Japan), SiC (particle size 2–3 μm, Kojundo Chemical Laboratory Co., Ltd., Japan), Si_3N_4 (particle size 1 μm, Kojundo Chemical Laboratory Co., Ltd., Japan), and α-Al_2O_3 (particle size 1 μm, Kojundo Chemical Laboratory Co., Ltd., Japan). Figures 1, 2, and 3 respectively show SEM micrographs

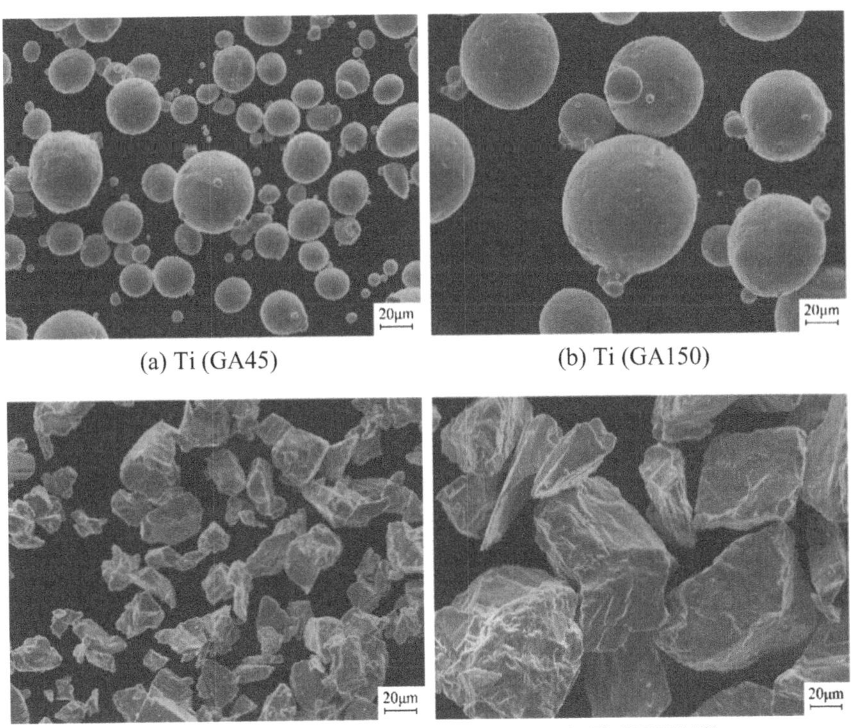

(a) Ti (GA45) (b) Ti (GA150)

(c) Ti (HDH45) (d) Ti (HDH150)

Fig. 1 SEM micrographs of pure titanium matrix powders

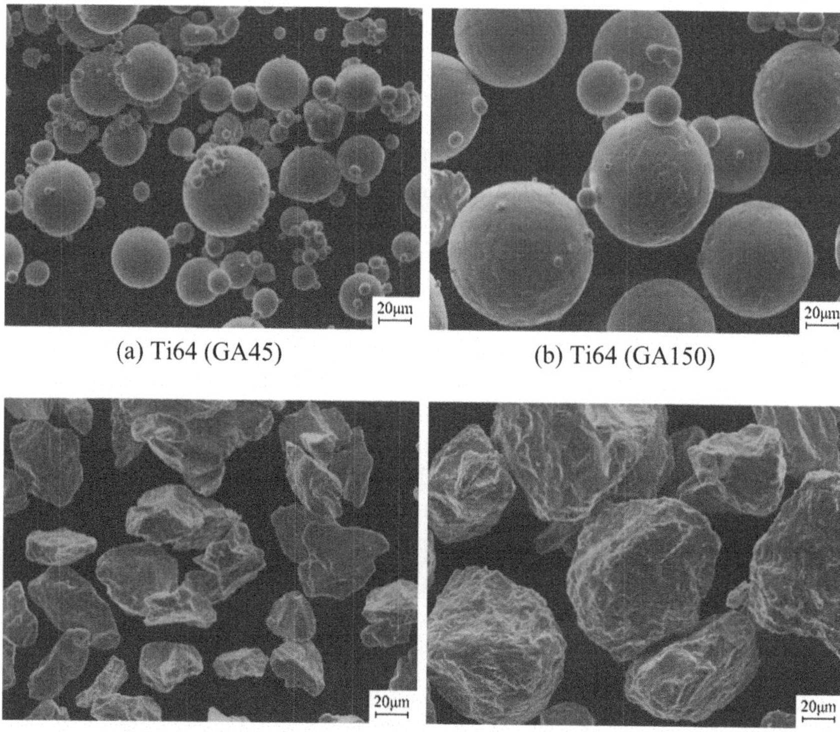

(a) Ti64 (GA45) (b) Ti64 (GA150)

(c) Ti64 (HDH45) (d) Ti64 (HDH150)

Fig. 2 SEM micrographs of Ti–6Al–4V titanium alloy matrix powders

of pure titanium matrix powders, Ti – 6Al – 4V titanium alloy matrix powders, and reinforcement powders, observed by a scanning electron microscope (SEM; SSX-550, Shimadzu Corporation, Japan).

The powder mixtures were blended using a planetary ball mill (P-6, Fritsch Japan Co., Ltd., Japan) at a rotating speed of 200 rpm for 10 min in air. The mixed powders were filled into a graphite die (Toyo Tanso Co., Ltd., Japan), compressed to a pressure of 200 MPa by a hand press, and then sintered by SPS (DR.SINTER SPS-3.20MK IV, Sumitomo Coal Mining Co., Ltd., Japan) after being evacuated to below 10 Pa. Sintering was performed at a sintering temperature of 1173 K, a heating rate of 293 K/min, a sintering time of 10 min, and a pressure of 70 MPa, followed by furnace cooling.

Characterization Tests

The composites were polished using 120-, 240-, 400-, 600-, 800-, 1200- and 1500-grit SiC emery paper until the average surface roughnesses (Ra) of the

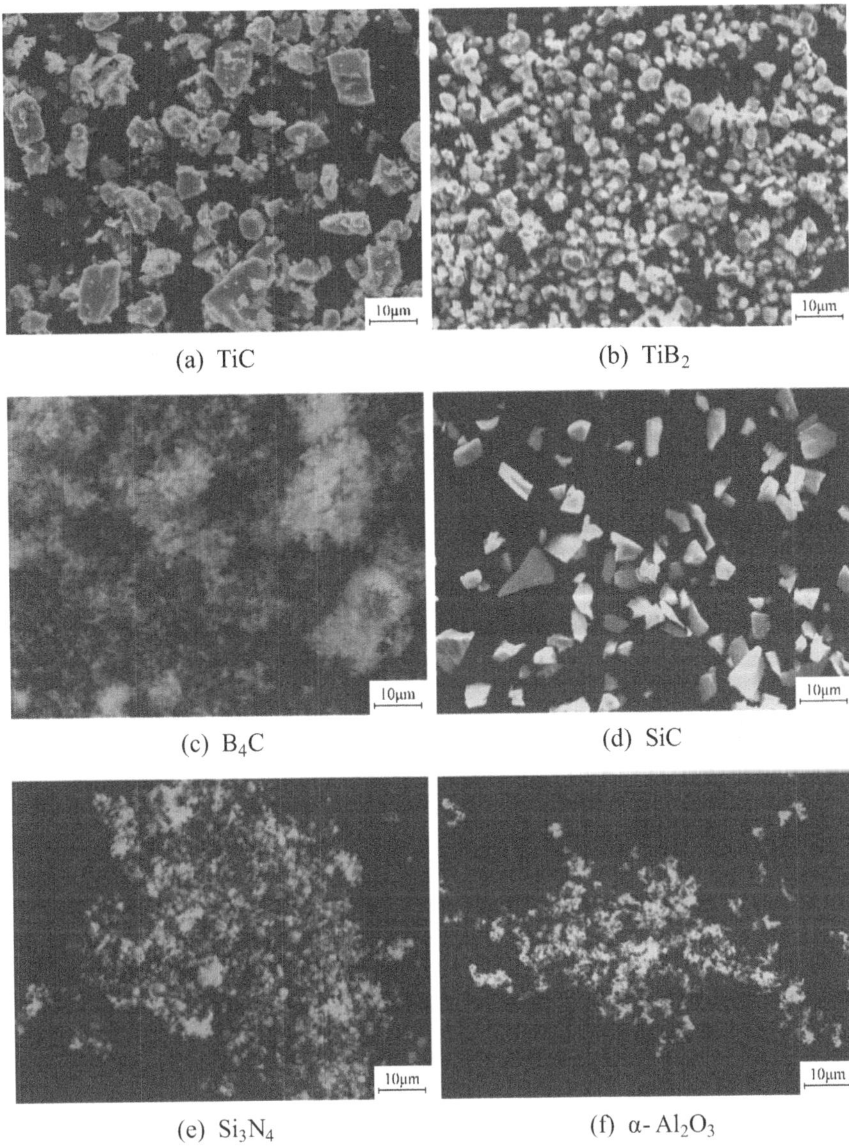

Fig. 3 SEM micrographs of reinforcement powders

composites were less than 0.2 μm. Then, the densities of the composites were measured by Archimedes' method using an electric balance (BX-420H, Shimadzu Corporation, Japan). Unlubricated (dry) sliding wear tests were carried out on a three-ball-on-disk wear test machine (MODEL EMF-3-F-ADX, A&D Company,

Fig. 4 Dry sliding wear tester

Limited, Japan), using steel ball bearings having a diameter of 10 mm and an average hardness of 810 HV, and the prepared composites serving as disks. The disks used in the test had a square shape with dimensions 35 × 35 mm and 4 mm in thickness. The three balls were placed at 120° intervals on a circle with a diameter of 28 mm. All wear tests were performed in air with the following parameters: a load of 40 N; a sliding speed of 100 mm/s; and a sliding distance of 500 m. The composites were washed with acetone for 5 min before and after the test, and the mass was measured using an electronic balance with a measurement range of 0.1 mg (HR-100A, A&D Company, Limited, Japan). Based on the obtained mass change and the measured density of the composite, a specific wear rate was calculated. Figure 4 shows the dry sliding wear tester.

Allocation to Orthogonal Array

In this study, five control factors that affected the wear resistance of the titanium matrix composites, as well as their levels, were chosen. Table 1 shows the control factors and their levels. When TiB_2 and B_4C were added to pure titanium or titanium alloy as the reinforcement material shown in Table 1, it was assumed that TiB and TiC + TiB were completely formed by reacting with Ti. In order to investigate the influence of the shape of the matrix in detail, the third level in the shape of the matrix was also set to "spherical". Similarly, the third level of the

Table 1 Control factors and their levels

Level	Control factor				
	A	B	C	D	E
	Reinforcement material	Reinforcement volume fraction	Matrix material	Shape of matrix	Particle size of matrix
1	TiC	5 vol.%	Pure-Ti	Spherical	<45 μm
2	TiB	10 vol.%	Ti–6Al–4V	Angular	<150 μm
3	TiC + TiB	15 vol.%	Pure-Ti+Ti–6Al–4V	Spherical	<150 μm
4	SiC	–	–	–	–
5	Si_3N_4	–	–	–	–
6	α-Al_2O_3	–	–	–	–

particle size of matrix was "<150 μm". An L_{18} orthogonal array in the Taguchi method was adopted because of the feature that the interaction among the elements of each column is uniformly scattered in the remaining columns [7]. These factors were allocated to the L_{18} orthogonal array, then the experiments were carried out according to this orthogonal array.

Results and Discussion

Table 2 shows the eighteen-orthogonal array (L_{18}) with results. From Table 2, the specific wear rate of No. 17 in the orthogonal array was the smallest. The specific wear rate $W_s[\mathrm{m}^2/\mathrm{N}]$ was calculated as follows:

$$W_s = \Delta W/PL\rho \tag{1}$$

where ΔW is the weight difference of the disk before and after the test, P is the normal load, L is the sliding distance, and ρ is the density of the disk.

Determination of optimum condition. The SN ratio η of the smaller-is-better response was calculated from the first and second specific wear rate as follows:

$$\eta = -10\log_{10}\left(\frac{W_{s1}^2 + W_{s2}^2}{2}\right) \tag{2}$$

where W_{s1} is the first specific wear rate, and W_{s2} is the second specific wear rate [8]. The level average was calculated based on the obtained SN ratio. Table 3 shows the level average of the SN ratio results. Then, a diagram of factorial effects related to the level average of the SN ratio was prepared, as shown in Fig. 5. Since the SN ratio in this study means that the larger the SN ratio, the better the wear resistance, the optimum condition for each factor is the level with a high SN ratio. Table 4 shows the optimum conditions for maximizing the wear resistance.

Table 2 Experimental design using $\mathbf{L}_{18}$ orthogonal array: specific wear rates and calculated S/N ratios for specific wear rates

Exp. No.	A	B	C	D	E	Specific wear rate 1	Specific wear rate 2	Output average	SN ratio
						W_{s1} [10^{-13}m^2/N]	W_{s2} [10^{-13}m^2/N]	$\overline{W_s}$ [10^{-13}m^2/N	η [dB]
1	1	1	1	1	1	4.3427	5.2587	4.8007	246.33
2	1	2	2	2	2	6.0515	7.0386	6.5451	243.66
3	1	3	3	3	3	8.4740	9.5967	9.0354	240.86
4	2	1	1	2	2	5.0767	4.8455	4.9611	246.09
5	2	2	2	3	3	6.4068	6.0888	6.2478	244.08
6	2	3	3	1	1	7.3774	7.3326	7.3550	242.67
7	3	1	2	1	3	6.0034	5.9888	5.9961	244.44
8	3	2	3	2	1	4.9904	5.6206	5.3055	245.49
9	3	3	1	3	2	3.9850	4.6017	4.2934	247.32
10	4	1	3	3	2	5.9000	6.4170	6.1585	244.20
11	4	2	1	1	3	3.4384	5.1462	4.2923	247.18
12	4	3	2	2	1	6.7539	5.1667	5.9603	244.42
13	5	1	2	3	1	3.6543	3.6258	3.6401	248.78
14	5	2	3	1	2	2.4842	3.4116	2.9479	250.50
15	5	3	1	2	3	3.4849	2.4653	2.9751	250.40
16	6	1	3	2	3	3.0555	3.0993	3.0774	250.24
17	6	2	1	3	1	2.1749	2.1868	2.1809	253.23
18	6	3	2	1	2	3.1537	7.2367	5.1952	245.06

Table 3 Level average of SN ratio results

Control factors	Levels	Level average of SN ratio $\bar{\eta}$ [dB]
A	1	243.62
	2	244.28
	3	245.75
	4	245.27
	5	249.90
	6	249.51
B	1	246.68
	2	247.36
	3	245.12
C	1	248.43
	2	245.07
	3	245.66
D	1	246.22
	2	246.72
E	1	246.82
	2	246.17

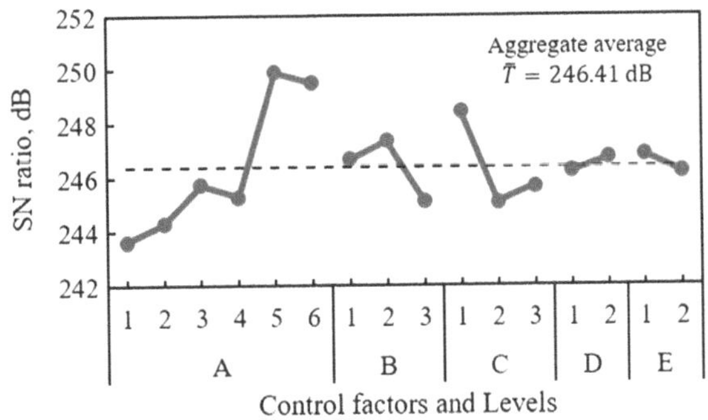

Fig. 5 Diagram of factorial effects

Table 4 Optimum conditions

Control factors	Level
Reinforcement material	Si_3N_4
Reinforcement volume fraction	10 vol.%
Matrix material	Pure-Ti
Shape of matrix	Angular
Particle size of matrix	$<45\mu m$

Other experiments were conducted to confirm whether the wear resistance with the determined optimum conditions was the most improved compared to all the tests in the orthogonal array. In these experiments, two test specimens with the optimum conditions shown in Table 4 were prepared, and two wear tests were carried out. Table 5 shows the results of the wear tests compared with No. 17 in the orthogonal array. From Table 5, it was confirmed that the wear resistance of the optimum condition was the most improved by the wear tests.

Prediction and Confirmation. For prediction and confirmation, an evaluation was conducted as to whether the experiments in this study were reproducible or not. In this evaluation, generally, the gain is obtained from the estimated value and the experimental value of the SN ratio under the optimum condition and the reference condition, and if the difference between the gain of the estimated value and the gain of the experimental value is ± 3.0 dB, it is considered that the experiments are reproducible [7, 8]. The estimated values of the SN ratio $\hat{\eta}$ under the optimum condition and the reference condition were calculated as follows:

Table 5 Results of wear tests at optimum condition compared with No. 17

	Specific wear rate 1 $W_{s1}[10^{-13}m^2/N]$	Specific wear rate 2 $W_{s2}[10^{-13}m^2/N]$	Output average $\overline{W}_s[10^{-13}m^2/N]$	SN ratio η [dB]
Optimum condition	1.6875	1.2307	1.4591	256.61
No. 17	2.1749	2.1868	2.1809	253.23

$$\hat{\eta} = \overline{T} + (A_i - \overline{T}) + (B_j - \overline{T}) + (C_k - \overline{T}) + (D_l - \overline{T}) + (E_m - \overline{T}) \quad (3)$$

where $A_i \sim E_m$ are the level average of the SN ratio in each control factor, $\overline{T}$ is the aggregate average of them, and subscripts $i \sim m$ are the level of each control factor. First, the estimated value of the optimum conditions was calculated. The optimum condition was $10Si_3N_4/Ti$(HDH45), the reinforcement material was the fifth level, the reinforcement volume fraction was the second level, the matrix material was the first level, the shape of matrix was the second level, and particle size of the matrix was the first level. Thus, the estimated value of the SN ratio under the optimum condition was as follows:

$$\begin{aligned}\hat{\eta} &= \overline{T} + (A_5 - \overline{T}) + (B_2 - \overline{T}) + (C_1 - \overline{T}) + (D_2 - \overline{T}) + (E_1 - \overline{T}) \\ &= A_5 + B_2 + C_1 + D_2 + E_1 - 4\overline{T} \\ &= 249.90 + 247.36 + 248.43 + 246.72 + 246.82 - 4 \times 246.41 \\ &= 253.59\ [\text{dB}]\end{aligned} \quad (4)$$

The reference condition was set to $15TiC/Ti$(GA45), which is the condition that was found to improve the wear resistance [6]. Therefore, the reinforcement material was the first level, the reinforcement volume fraction was the third level, the matrix material was the first level, the shape of the matrix was the first level, and the particle size of the matrix was the first level. Thus, the estimated value of the SN ratio under the reference condition was as follows.

$$\begin{aligned}\hat{\eta} &= \overline{T} + (A_1 - \overline{T}) + (B_3 - \overline{T}) + (C_1 - \overline{T}) + (D_1 - \overline{T}) + (E_1 - \overline{T}) \\ &= A_1 + B_3 + C_1 + D_1 + E_1 - 4\overline{T} \\ &= 243.62 + 245.12 + 248.43 + 246.22 + 246.82 - 4 \times 246.41 \\ &= 244.57\ [\text{dB}]\end{aligned} \quad (5)$$

Table 6 shows the results of evaluating the reproducibility. From Table 6, since the gain difference between the estimated value and the experimental value in these experiments was ±2.77 dB, which is within the range of ±3.0 dB, it is considered to be reproducible.

Influence of control factors. Three control factors influenced the wear resistance of the titanium matrix composites: reinforcement material, reinforcement volume fraction, and matrix material. Among the factors that had the main effect, the reinforcement volume fraction was investigated. To investigate the reason why the optimum level of the reinforcement volume fraction became 10vol.%, two test specimens under two conditions ($5Si_3N_4/Ti$(HDH45), $15Si_3N_4/Ti$(HDH45)) were prepared, in which only the reinforcement volume fraction was changed, in addition to the optimum condition ($10Si_3N_4/Ti$(HDH45)). Then, wear tests of these specimens were carried out. Figure 6 shows the results of the wear tests. Vickers microhardness tests of these specimens were carried out using a microhardness

Table 6 Prediction and Confirmation

	SN ratio [dB]	
	Estimated value	Experimental value
Optimum condition $10Si_3N_4/Ti$(HDH45)	253.59	256.61
Reference condition 15TiC/Ti(GA45)	244.57	244.82
gain	9.020	11.79

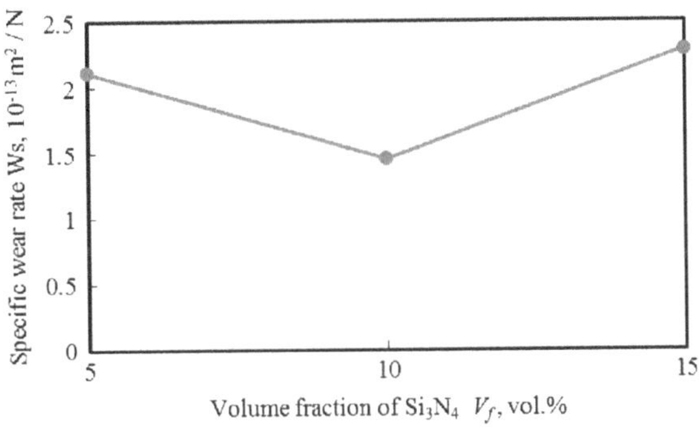

Fig. 6 Specific wear rate versus volume fraction of Si_3N_4 of the composites

tester (HMV-2(T), Shimadzu Corporation, Japan). Figure 7 shows the results of the Vickers microhardness tests. From Fig. 7, the Vickers hardness increased as the volume fraction of Si_3N_4 (V_f) increased. Elemental analysis was carried out by energy dispersive X-ray spectrometry (EDX; SUPERSCAN SSX-550, Shimadzu Corporation, Japan). Figure 8 shows EDX mapping images of $10Si_3N_4/$ Ti(HDH45). Analysis of the structural component was performed by X-ray diffraction (XRD; RINT2000X, Rigaku Co., Ltd., Japan). Figure 9 shows XRD patterns of Si_3N_4/Ti(HDH45) composites. From Fig. 8, there were three regions in the microstructure of the composites. In region I shown in Fig. 8, titanium and nitrogen are distributed. In region II, mainly silicon is distributed, and small

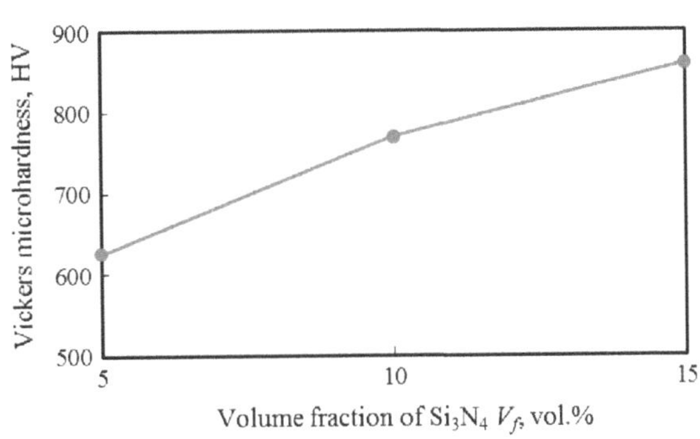

Fig. 7 Vickers microhardness versus volume fraction of Si_3N_4 of the composites

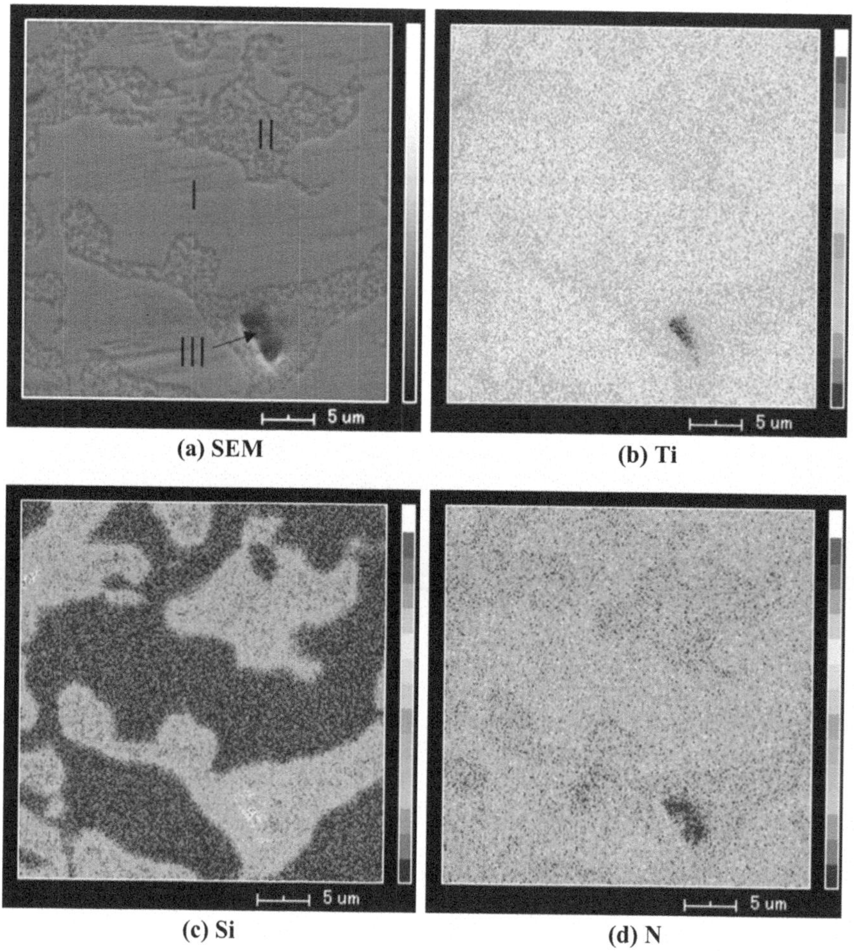

Fig. 8 EDX-mapping images of $10\mathbf{Si_3N_4/Ti}(\mathbf{HDH}45)$

amounts of titanium and nitrogen are distributed. In region III, silicon is distributed at a higher concentration than in region II. From (b) and (d) in Fig. 8 titanium and nitrogen are similarly distributed. From the X-ray diffraction results shown in Fig. 9, the reaction products Ti_5Si_3 and TiN were detected. Considering the standard Gibbs free energy of the reaction, the chemical reaction formula between Ti and Si_3N_4 was obtained as follows [9, 10]:

$$\frac{3}{2}\mathrm{Si} + \mathrm{N_2} = \frac{1}{2}\mathrm{Si_3N_4} \quad \Delta G^0 = -373.1 + 0.166T \tag{6}$$

$$2\mathrm{Ti} + \mathrm{N_2} = 2\mathrm{TiN} \quad \Delta G^0 = -673.3 + 0.187T \tag{7}$$

$$5Ti + 3Si = Ti_5Si_3 \quad \Delta G^0 = -579.8 + 0.120T \tag{8}$$

$$9Ti + Si_3N_4 = 4TiN + Ti_5Si_3 \quad \Delta G^0 = -1180.2 + 0.162T \tag{9}$$

where $\Delta G^0[\mathrm{kJ/mol}]$ is the difference of the standard Gibbs free energy of the reaction, and $T[\mathrm{K}]$ is the temperature. From formula (9), at the sintering temperature ($T = 1173K$), ΔG^0 is $-990.2\mathrm{kJ/mol}$, which is a negative value, so it is considered that when Si_3N_4 is added to Ti, TiN and Ti_5Si_3 are reactively formed. Therefore, it is considered that TiN is uniformly dispersed, whereas Ti_5Si_3 is dispersed in region II. As shown in Fig. 9, the diffraction peak of α-titanium moved to the lower angle side as V_f increased. This is probably because the nitrogen atom which is an interstitial solid solution is dissolved in the crystal lattice of α-titanium [11]. Therefore, it is considered that dispersion of Ti_5Si_3 and TiN and solid solution strengthening of nitrogen atoms increased the Vickers hardness of the composites. As shown in Fig. 6, the specific wear rate was the smallest when V_f was 10 vol.%. To investigate the cause, the cross sections of the wear tracks in the composites were observed using an optical microscope (DSX510, Olympus Corporation, Japan). Figure 10 shows the observed cross sections of a wear track. The presence of a crack was confirmed, as shown in Fig. 10, which was most frequently observed when V_f was 15 vol.%. It is considered that at 15 vol.%, embrittlement progressed due to an increase in the amount of brittle Ti_5Si_3 produced compared with 5 vol.% and 10 vol.%. Therefore, the surface of the disk was more damaged and worn during sliding as V_f increased [12]. From the above, when Si_3N_4 was added to pure titanium, solid solution strengthening by nitrogen atoms and dispersion products increased the Vickers hardness and improved the wear resistance. However, due to the embrittling as V_f increased, cracking was likely to occur on the surface of the disk, the wear phenomenon in which the surface was noticeably damaged appeared at 15 vol.%, and the wear resistance was affected. Accordingly, it is considered that the wear resistance was most improved when V_f was 10 vol.%.

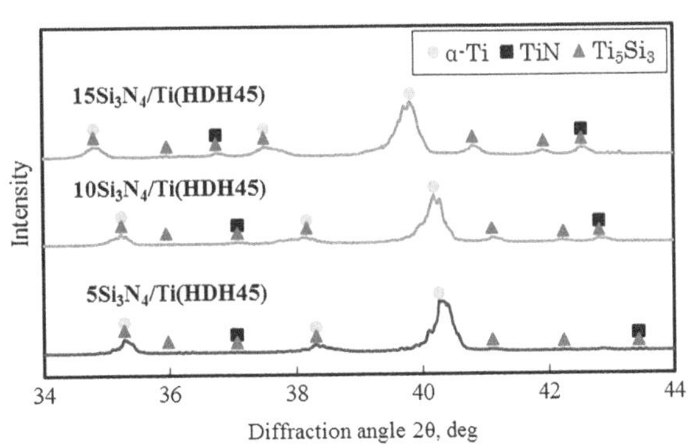

Fig. 9 XRD patterns of Si_3N_4/**Ti**(**HDH**45) composites

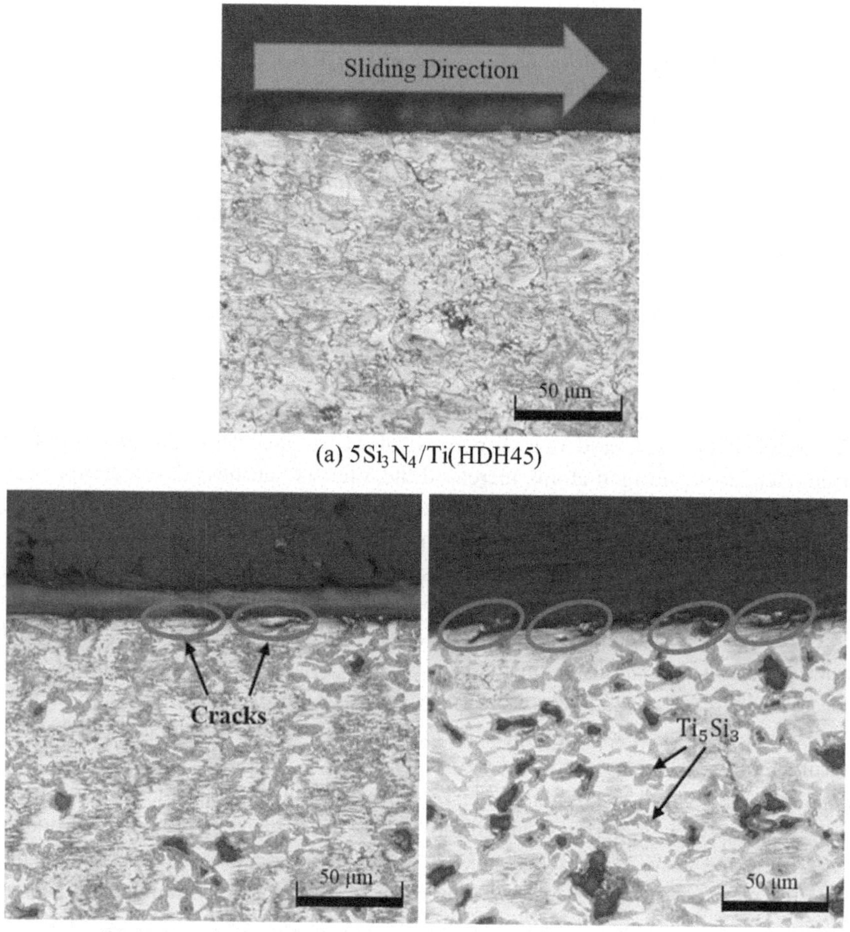

(a) $5Si_3N_4$/Ti(HDH45)

(b) $10Si_3N_4$/Ti(HDH45)

(c) $15Si_3N_4$/Ti(HDH45)

Fig. 10 Cross sections of wear tracks in Si_3N_4/**Ti**(**HDH**45) composites

Summary

The results obtained in this work can be summarized as follows:

1. The main control factors affecting the wear resistance of the titanium matrix composites were the reinforcement material, the reinforcement volume fraction, and the matrix material. On the other hand, the shape of the matrix and the particle size of the matrix had little influence.
2. The Si_3N_4 reinforced hydride–dehydride Ti matrix composite with 10 vol.% reinforcement showed good wear resistance (optimum condition: $10Si_3N_4$/Ti(HDH45)).

3. When Si_3N_4 was added to pure titanium, solid solution strengthening by nitrogen atoms and reaction products Ti_5Si_3 and TiN increased the hardness and improved the wear resistance.
4. The wear resistance of $15Si_3N_4/Ti(HDH45)$ with excessive addition of Si_3N_4 was likely to crack on the surface due to the formation of brittle Ti_5Si_3, and cause the wear phenomenon where the surface was more noticeably damaged.

In this study, no noise factors were taken into consideration; however, in Taguchi methods, experiments taking noise factors into consideration are indispensable for developing robust technologies. Thus, from now on, we plan to study appropriate noise factors and perform functional evaluations in experiments including noise factors.

References

1. Inagaki I, Takechi T, Shirai Y, Ariyasu N (2013) Application and features of titanium for aerospace industry. Nippon Steel Sumitomo Met Tech Rep 396:23–28
2. Choda T (2015) Technology for process design of titanium alloy forging. J Jpn Inst Light Met 65(9):460–465
3. Basavarajappa S, Chandramohan G, Davim JP (2007) Mater Des 28:1393–1398
4. Prakash U,Ajith Prasad SL,Ravindra HV (2015) Mater Today Proc 2:1825–1832
5. Yamamoto Y, Kaneta M (2014) Tribology, 2nd edn, Ohmsha Co. Ltd.
6. Kamegawa S, Izui H, Komiya Y (2016) Tensile and wear properties of TiB/Ti and TiC/Ti composites with different Ti powders prepared by spark plasma sintering. Master's thesis of Nihon University
7. Hasegawa Y (2013) The approach to solve environmental problems using Taguchi Methods. JUSE Press, pp 66, 72
8. Tatebayashi K (2015) Introduction to taguchi methods. JUSE Press, pp 27–29, 116,117
9. Onzawa T, Suzumura A, Kim JH (1990) Filament/Matrix Interaction of Titanium-Matrix Composites Reinforced with SiC Fibers. Q J Jpn weld Soc 8(2):286–292
10. The Japan Institute of Metals: Physical Chemistry of Metals, (2013), vol 1. pp 208, 209
11. Yamabe Y, Umeda J, Imai H, Kondoh K (2017) J Jpn Soc Powder Metall 64(6):275–280
12. Liu Yuan-Fu, Zhou Yong-Lin, Zhang Qian, Fei Pu, Li Rong-Hua, Yang SZ (2014) J Alloy Compd 591:251–258

Model-Based Damage Detection in Piezoelectric Fiber Based Composites

Khalid M. Shalan, Mohamed E. AbdelMeguid, Tarek M. Hatem, Hesham A. Hegazi and Yehia A. Bahei-El-Din

Abstract Piezoelectric Fiber-based Composites (PFCs) have significant potential as smart materials given their superior mechanical properties over piezoelectric wafers. Therefore, reliable models are needed to accurately predict PFCs behavior including inherent heterogeneity and coupled electro-mechanical fields. This paper offers a multi-resolution model (micro and macro) that calculates the homogenized moduli of heterogeneous PFCs including the coupled electro-mechanical field based on Transformation Field Analysis (TFA). The calculated properties from TFA micro analysis is used in a macro-scale finite element analysis to model the dynamic behavior of PFCs in macro-scale. Numerical dynamic analysis incorporates a simple structure in pristine condition and in two damage conditions, namely a delamination damage and an impact-induced damage, modeled using both modal analysis and implicit dynamic analysis.

Keywords Structural health monitoring · Piezoelectric fiber composites Modal analysis · Delamination · Transformation field analysis

---

K. M. Shalan (✉) · M. E. AbdelMeguid · T. M. Hatem (✉) · Y. A. Bahei-El-Din
Centre for Simulation Innovation and Advanced Manufacturing,
The British University in Egypt, 11837 El-Sherouk City, Cairo, Egypt
e-mail: Khalid.Shalan@bue.edu.eg

T. M. Hatem
e-mail: tarek.hatem@bue.edu.eg

M. E. AbdelMeguid
e-mail: mohamed_ezz_nasr@hotmail.com

Y. A. Bahei-El-Din
e-mail: ybahei@bue.edu.eg

K. M. Shalan · H. A. Hegazi
Mechanical Design and Production Department, Faculty of Engineering,
Cairo University, 12613 Giza, Egypt
e-mail: Hhegazi@aucegypt.edu

T. M. Hatem
Microstructure Physics and Alloy Design Department, Max-Planck-Institut
Fur Eisenforschung, 40237 Düsseldorf, Germany

T. S. Srivatsan et al. (eds.), *Metal-Matrix Composites Innovations, Advances and Applications*, The Minerals, Metals & Materials Series,
https://doi.org/10.1007/978-3-319-72853-7_9

Introduction

The piezoelectric materials capability of reversible coupling between mechanical and electrical fields is a key to their usage as smart materials [1–3]. Piezoceramics materials have been widely used as both an actuator and a sensor in a wide range of applications given their wide dynamic range [4, 5]. In addition, piezoelectric materials can act as a small power source harvesting ambient vibrations in electrical energy [6]. However piezoceramics are limited by their inherent brittleness as a ceramic materials, in addition to high density causing impedance mismatch with lower density composites, live tissues and water [7, 8]. Piezopolymers on the other hand promise better mechanical properties on the expense of the piezoelectric capabilities in comparison to piezoceramics [9]. Piezoelectric composites made up of piezoceramics embedded in polymer matrix offered tailor-made properties in addition to increased toughness and compliance and lowered density and dielectric constant. Piezoelectric composites (piezocomposites) can be classified according to structure and connectivity of each constituent (0, 1, 2, or 3 dimensionality).

Hagood and Bent [10] introduced Active Fiber Composites (AFCs) constituting Lead-Zirconate-Titanate (PZT) piezoceramics fibers embedded in epoxy matrix. Wilkey et al. at NASA Langley Research Center introduced Macro Fiber Composites (MFCs) [11, 12] where rectangular PZT rods from a sliced wafer were used to lower the cost of manufacturing. In both cases, Interdigitated Electrodes (IDE) are implemented to utilize the higher piezoelectric d33 and e33 effects (along fiber direction) as d33 is usually around 2.4 times higher than the d31 effect and e33 can be 4 times higher [13]. Interdigitated electrodes, first introduced by White and Voltmer [14], were implemented in AFCs and MFCs. IDEs consist of comb shaped electrode whose teeth are perpendicular to the fibers on its top and bottom. The other pole electrode would be added from the other side to occupy the interstitials of the first (without touching the first electrode, elementarily). Achieving a uniform response entails the PFC being poled after manufacturing [15].

PFCSs show superior mechanical properties [16, 17] along with their conformity to curvature [18]. Many studies showed great potential of PFCs in energy harvesting applications [19, 20], Structural Actuation [10, 21–23], Vibration control [24] as well as Structural Health Monitoring applications [25, 26].

Modelling PFCs includes the inherent complexities of modelling composite materials to the coupled electric and mechanical fields. Many attempts were made to provide analytical or numerical models that could predict the properties of active composites. Solutions depended on expanding inactive composite materials models starting from the early Eshelby's infinite matrix with ellipsoidal inclusion, further expanded by Mori-Tanaka [27, 28]. Dunn and Taya used a Mori-Tanaka based model to evaluate the effective elastic moduli of a piezoelectric composite [29], while Li and Dunn expanded the approach to account for coupling between mechanical, electrical and magnetic fields [30]. Exact solutions for the relation between local fields and effective moduli in piezoelectric fiber composites were attempted by Benvensite [31]. Schulgasser approached piezoelectric fiber composites as transversely isotropic material and studied relations between overall properties [32].

Asymptotic expansion homogenization techniques extrapolate the electro-mechanical behavior of electrically active composite using a unit cell model to characterize an idealized periodic geometry through the governing equations of local fields [33]. Bahei-El-Din expanded on this approach to compute the overall electro-mechanical response of electrically active woven composites using Transformation Field Analysis (TFA) while accounting for damage in composite lamina [34]. The latter model was augmented with Carrera Unified Formulation [26] and macroscale finite element analysis [19, 35].

One of the most prominent techniques used in structural health monitoring are the modal analysis techniques where mechanical excitation is introduced and the natural frequencies of the structure are detected by sensors. The detection of any shifts in the natural frequencies than the pristine case is used to identify damage [36, 37]. Piezoelectric materials can perform the role of both an actuator introducing mechanical excitation at and a sensor converting the displacements into an electric signal that be connected to directly to data acquisition instrumentation [35].

In this paper, the active material properties of PFC composites used to describe the composite overall electromechanical behavior are obtained from the TFA model. Accordingly, the PFC composite is treated as a homogenous transversely isotropic material with coupled electro-mechanical behavior. The resulting homogenized properties are used in Finite Element analysis performed on commercial package (Abaqus). The multiscale analysis is performed on Piezoelectric Fiber Composites Bimorph (PFCB-14) manufactured by Advanced Cerametrics Inc. [38] as an example of a simple structure incorporating PFCs. PFCB consists of two PFC layers bonded to a stainless-steel sheet, which enhance the rigidity of the structure. Pristine samples are delaminated between the PFC layer and steel sheet to quantify the difference in the dynamic response through finite element analysis.

Advancements in modelling for PFC would entail wider usage especially in Structural Health Monitoring (SHM) applications. The variables to be used for detection in neural networks analysis, known as Feature Vector elements, can be thus estimated using numerical calculations [39]. Model-informed SHM systems can have database of results for supervised training and prognostics from the numerical studies to evaluate damage [37].

Micromechanics Study

Constitutive Model

Piezoelectric response can be defined through two reversible effects; first, the production of an electric field as a direct response to an applied stress; and second, a converse effect represented by mechanical deformation as a response to an applied electric potential. The linearly formulated constitutive equations binding the response of piezoelectric material are standardized by IEEE [40]. The constitutive equations for piezoelectric material are derived from the thermo-dynamical principles correlating the electric field [*E*], strain [*S*], stress [*T*] and electrical displacement [*D*]

$$S = s^E T + dE \tag{1}$$

$$D = d^T T + \varepsilon E_k \tag{2}$$

Transformation Field Analysis

TFA defines Eigen stresses $[\lambda]$ and strains $[\mu]$ for the transformation fields in the RVE, the case shown being due to piezoelectric effects. The constitutive equations for each element are written to include induced transformation fields due to uniform stress $[T_i]$ or strain $[S_i]$.

$$S_i = s_i^E T_i + \mu_i \tag{3}$$

$$T_i = C_i^E S_i + \lambda_i \tag{4}$$

where i represent number of elements inside the RVE [1, 2 … Q]. Comparing equations with constitutive equations for PFC, the transformation fields can be attributed to an applied electric field in electrically active piezoelectric fiber-based composites, using Eqs. 2, 4, and 5.

$$\mu_i = d_i E_i \tag{5}$$

$$\lambda_i = -e_i E_i \tag{6}$$

For the RVE entity, the strains and stress caused by uniform stress or strain are super-positioned across the entire volume.

$$S_i = A_i S + \sum\nolimits_{j=1,\Omega} D_{ij} \mu_j \tag{7}$$

$$T_i = B_i T + \sum\nolimits_{j=1,\Omega} F_{ij} \lambda_j \tag{8}$$

where, Ω is the number of elements carrying transformation fields inside the RVE, the piezoelectric fibers. A_i and B_i are concentration factors used to describe the volume strain and stress in terms of overall counterparts. D_{ij} and F_{ij} are constant influence functions depends mainly on the elastic moduli of each element.

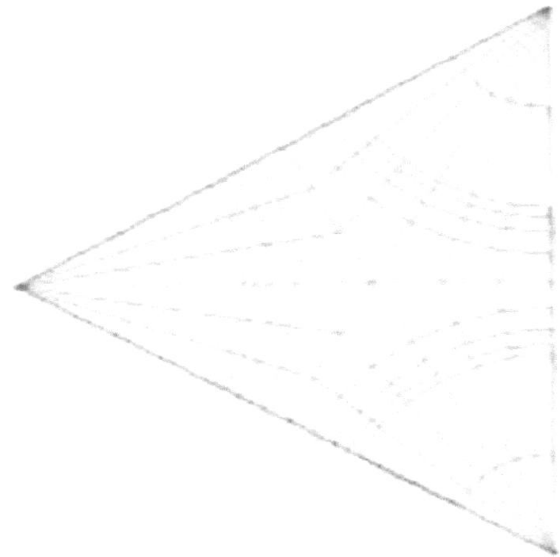

Fig. 1 PHA representative volume element

Table 1 Material properties of PZT-5A and polymer [41]

	E_L	ν_L	d_{33}	d_{31}	d_{15}	ε_{33}
	(GPa)		$\left(\frac{m}{V}x10^{-12}\right)$			$\left(\frac{C}{V.m}x10^{-9}\right)$
PZT 5-A	69	0.31	374	−171	584	15,000
Polymer	3.35	0.35	–	–	–	–

These matrices are numerically attained through finite element analysis of the RVE under potential difference along the 3-3 direction (fiber direction) at different voltages values.

The micromechanics study was performed on a Periodic Hexagonal Array (PHA), that is a representative volume element of triangular shape with periodic boundaries to represent hexagonal packing of fibers. 300 elements were used to represent the piezoceramic Lead Zirconate Titanate (PZT) fiber, the epoxy matrix and the interface between them at a higher resolution, as shown in Fig. 1. Boundary condition of voltage is applied along the fiber direction and the mechanical displacement is measured in the same direction for alternating voltages.

The TFA solution was considered for PFC composite consisting of two constituents PZT-5A and Polymer resin, the material properties of each is described in Table 1.

Homogenized Material Properties

A comparison is held for the TFA results with the manufacturer tabulated data as shown in Fig. 2, the TFA results present purely linear behavior based on the constitutive equations for piezoelectric material, while the experimental behavior

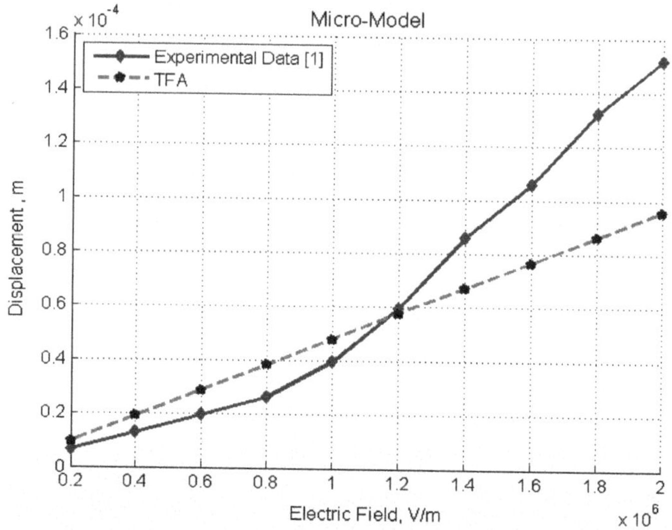

Fig. 2 Tabulated manufacturer data compared with TFA results for mechanical deformation under applied electrical load [38]

demonstrates a non-linear behavior [38]. Good accordance is demonstrated between the TFA model and experimental results on the low-medium voltage range while the results diverge as medium-high voltage is applied. Wide divergence occurs at very high voltages.

The PFC material could be thus represented as a homogenized transversely isotropic piezoelectric material with the following properties:

$$\textit{Elastic \& Shear Moduli pseudovector}: \quad E = \begin{bmatrix} 6E9 \\ 6E9 \\ 28.7E9 \\ 1.24E9 \\ 1.24E9 \\ 3.29E9 \end{bmatrix} Pa;$$

$$\textit{Electric Permittivity Matrix}: \quad \epsilon = \begin{bmatrix} 1.5E-8 & 0 & 0 \\ 0 & 1.5E-8 & 0 \\ 0 & 0 & 1.47E-8 \end{bmatrix} F/m$$

Piezoelectric Strain Tensor :

$$d = \begin{bmatrix} -1.625E-10 & 0 & 0 \\ 0 & -1.625E-10 & 0 \\ 0 & 0 & 3.5E\quad 10 \\ 0 & +5.640E-10 & 0 \\ +5.640E-10 & 0 & 0 \\ 0 & 0 & 0 \end{bmatrix} C/N$$

$$In-plane\,Poisson\,Ratio: \quad \nu_{12} = 0.31$$

Macroscale Finite Element Study

A Bimorph PFC (PFCB) consists of two layers of PFCs with a steel sheet in between. To represent such geometry, a layer of steel sheet of thickness 0.5 mm, width 14 mm and cantilevered length of 130 mm (equal to the active length of the fibers) is covered by strips of PFC homogenized material on both sides, each is 0.3 mm. The length used corresponds to the cantilever length from free end to support used in all numerical experiments. Built–in elements in Abaqus were used to simulate the model. Implicit Cubic (8 nodes) Linear 3D Stress Elements with reduced integration and hourglass stiffness control (C3D8R) were used to the steel lamina. Whereas the piezoelectric version of the aforementioned element type C3D8E were used for the piezoelectric materials. C3D8E adds a degree of freedom per node for the electric potential, and is only offered for direct-integration scheme (implicit analysis). One of the extreme faces along the longitudinal axis is used as an electric ground as well as fixed surface (encastre condition).

Following the approach by Birchmeier et al. [15] and to lessen the computational cost, the Interdigitated Electrodes (IDE) were replaced with electrodes at the extreme faces along the longitudinal/ fiber direction, subsequently voltages in model would have to be higher to create similar electric field.

Damage Modelling

Damage was introduced as a localized loss of stiffness for the material. In the analysis two cases of damage were studied: delamination damage and local failure due to impact. Delamination was introduced as a thin layer of material at 10% of the original elastic moduli values. The latter type of damage was introduced using a circular region of reduced stiffness extending through the three layers (10% of original elastic moduli at each layer of the circular damaged zone). The reduced stiffness approach has been demonstrated by Hatem et al. [37]. The delamination is inserted at 50 mm off the tip and extends for 28 mm. meanwhile, the impact induced damage is circular of radius 3 mm.

Modal Analysis

As piezoelectric materials couple the mechanical strain to a formed electrical field within the material; electric resonance couples the mechanical resonance causing a suddenly higher voltage output at the same natural frequencies. The phenomenon is referred to as electromechanical resonance where the piezoelectric material's impedance drops as well causing a higher voltage output. The piezoelectric electromechanical resonance is thus employed to simplify the modal analysis required to identify the dynamic properties of the bimorph in hand to a single input single output approach.

Eigen analysis using Lanczos solver was used on Abaqus to calculate the natural frequencies of the PFCB beam in cantilever position. The convergence behavior of the model was tested by running successive eigen-analyses and testing for the fundamental frequency. The test started by a large seed length of element (2 mm), which is less than thickness of the PFC layer. As results the length and width of the element remained longer than the thickness (i.e. larger than seed length of 0.3 mm) the results showed a converging behavior about 25 Hz. However, at seed length of 0.3 mm, the results diverged to about 31 Hz and rapidly converged, as shown in Fig. 3. The elements used had the size of 1/3 mm × 2 mm × 2 mm. The flexural behavior of the PFC layer is better captured with multiple elements across the thickness of each layer.

Dynamic Simulation

To idealize the motion of a PFCB under harmonic excitation by a shaker, a mechanical Displacement replaced the encastre condition in the earlier analysis. In other words, the upper and lower surface of PFCB at one end are gripped for a length of 10 mm by allowing only harmonic displacement along the vertical y-axis (referred to earlier as axis 1 in piezoelectric definitions) of amplitude of 1 mm at

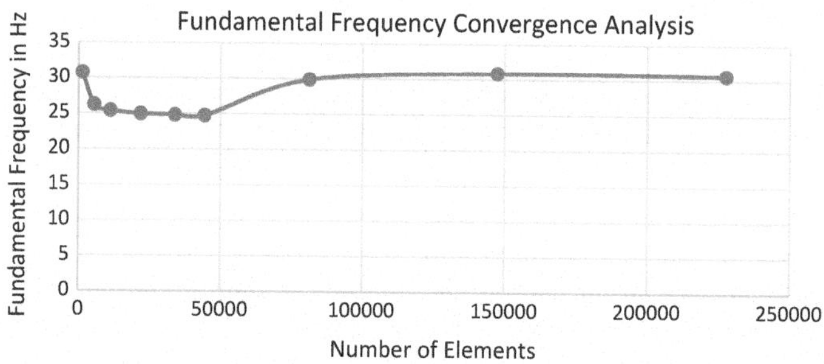

Fig. 3 Results for convergence analysis using 1st modal frequency of PFCB-W14

20 Hz. Inertial Loads due to PFCB mass were added as they would affect the flexural behavior of the shaken PFCB.

Results

The modal frequencies of the PFCB showed minor changes between the pristine sample and the delaminated sample, more significant with latter modes. The change became more significant in comparison with the model with impact-induced damage. The mode shapes remained insignificantly changed. The modes of the pristine sample can be shown in Fig. 4.

The shifting occurring due to damage can be summarized as follows in Table 2.

Sinusoidal mechanical excitation at a frequency of 20 Hz resulted in different electrical response at the electrodes of the sample. Given that one of the electrodes was defined to have zero voltage as a reference, the readings were taken at the other end of the PFCB, presented in Fig. 5.

The electrical response shows a super position of faster set of frequencies representing the free vibration response at the system's natural frequencies, given the system being defined without structural or Raleigh damping. Moreover the delamination damage shows a slight change in the electrical response while the response of the impact-induced damage goes in and out of phase from the pristine response.

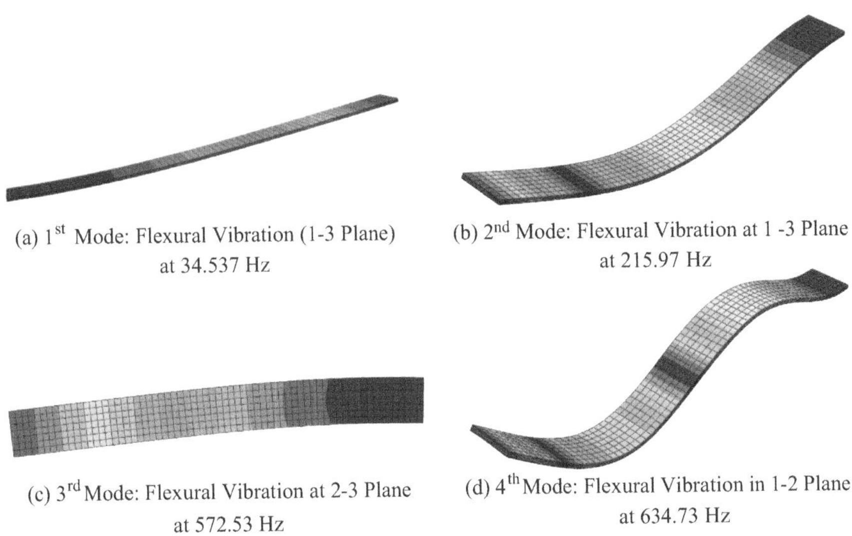

(a) 1st Mode: Flexural Vibration (1-3 Plane) at 34.537 Hz

(b) 2nd Mode: Flexural Vibration at 1 -3 Plane at 215.97 Hz

(c) 3rd Mode: Flexural Vibration at 2-3 Plane at 572.53 Hz

(d) 4th Mode: Flexural Vibration in 1-2 Plane at 634.73 Hz

Fig. 4 Mode shapes of a pristine PFCB

Table 2 Modal frequencies in pristine and damaged models

	Pristine Sample	Delaminated sample	Impact-damage sample
1st mode: Flexure 1–3 plane (Hz)	34.53	34.74	36.332
2nd mode: Flexure 1–3 plane	215.97	217.48	221.58
3rd mode: Flexure 2–3 plane	572.53	572.85	573.48
4th mode: Flexure 1–3 plane	608.57	614.43	634.73

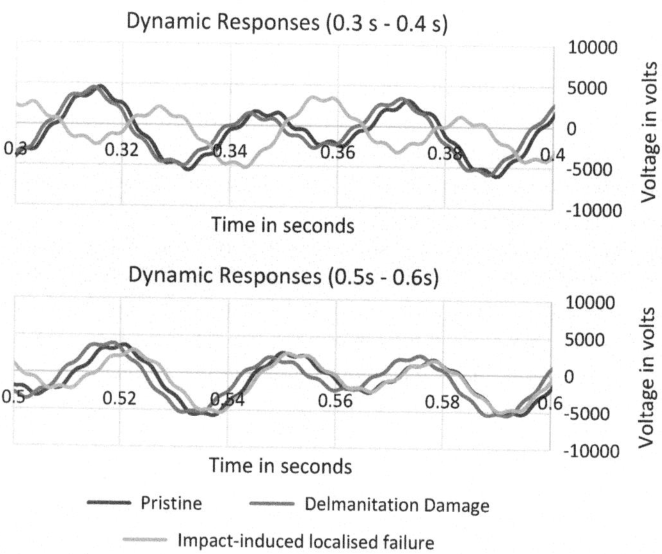

Fig. 5 Two segments of the dynamic response of the three samples due to mechanical flexural excitation at 20 Hz

Conclusions

The presented study provides a multiscale computational model for Piezoelectric Fiber Composite (PFC) for damage detection applications is Structural Health Monitoring. In a micromechanics study, Transformation field analysis was used to calculate overall effective electromechanical properties of the heterogenous piezoelectric fiber composites. Representing PFCs as a homogenized material contributes to lowering the computational cost of modelling PFC in macroscale studies. Moreover, the convenience of macroscale FE methods can model PFCs in a wide array of applications including structural health monitoring systems.

The results from the TFA method used, based on a linear constitutive model, show good accordance with the manufacturer tabulated data on low-medium voltage, suitable to the applications of sensing and wave generation commonly used in structural health monitoring.

While the modal analysis has the benefit of a low computational cost in modelling (using linear eigen analysis), its efficacy for damage detection has shown to be limited in the presented study. However, further studies are needed to evaluate the suitable ranges for modal analysis to be used in detection. More complex dynamic analyses, based on implicit dynamics, show more observable effect in the analysis results due to different types of damage. Thus, dynamic analysis can provide more modelling capability for more reliable damage diagnostics, especially if combined with proper signal processing techniques. Replicating the signal output in experimental and field setup would still need further studies especially of the damping characteristics of the PFC.

References

1. Ikeda T (1996) Fundamentals of piezoelectricity. Oxford University Press, UK
2. Bryant RG (2007) Overview of NASA Langley's piezoelectric ceramic packaging technology and applications: NASA technical report, NASA Langley research center. Available at https://ntrs.nasa.gov/archive/nasa/casi.ntrs.nasa.gov/20080000875.pdf
3. Cady WG (1946) Piezoelectricity. McGraw Hill, London
4. Kholkin AL et al (2008) Smart ferroelectric ceramics for transducer applications. In: Schwartz E (ed) Smart materials. CRC Press, Taylor & Francis Group, LLC, UK
5. Uchino K, Ito Y (2008) Smart ceramics: transducers, sensors, and actuators. In: Schwartz E (ed) Smart materials. CRC Press, Taylor & Francis Group, LLC, UK
6. Erturk A, Inman DJ (2011) Piezoelectric energy harvesting. John Wiley & Sons, USA
7. Moulson AJ, Herbert JM (2003) Electroceramics: materials, properties, applications. John Wiley & Sons, USA
8. Gandhi MV, Thompson B (1992) Smart materials and structures. Springer Science & Business Media, Germany
9. Ebnesajjad, S (2015) Fluoroplastics, Volume 2—Melt processible fluoropolymers—the definitive user's guide and data book. William Andrew Publishing
10. Hagood N, Bent A (1993) Development of piezoelectric fiber composites for structural actuation. In: 34th Structures, Structural Dynamics and Materials Conference, pp 3625–3638
11. Wilkie WK et al (2000) Low-cost piezocomposite actuator for structural control applications. In: SPIE's 7th annual international symposium on smart structures and materials. International Society for Optics and Photonics
12. Wilkie WK et al (2003) Method of fabricating a piezoelectric composite apparatus. Google Patents
13. Hagood NW et al (1993) Improving transverse actuation of piezoceramics using interdigitated surface electrodes. In: 1993 North American conference on smart structures and materials. International Society for Optics and Photonics
14. White R, Voltmer F (1965) Direct piezoelectric coupling to surface elastic waves. Appl Phys Lett 7(12):314–316
15. Birchmeier M et al (2009) Active fiber composites for the generation of Lamb waves. Ultrasonics 49(1):73–82
16. Bent AA (1997) Active fiber composites for structural actuation, Ph.D. Thesis. Massachusetts Institute of Technology, Cambridge, Massachusetts

17. Wilkie W, High J, Bockman J (2002) Reliability testing of NASA piezocomposite actuators. In: Proceedings of Actuator 2002: 8th International Conference on New Actuators. Bremen, Germany
18. Sodano HA, Park G, Inman DJ (2003) Vibration testing and control of an inflatable torus using multiple sensors/actuators. AIAA, AIAA Paper, p 1644
19. El-Etriby A et al (2014) A multiscale-based approach for composite materials with embedded PZT filaments for energy harvesting. In: smart structures and non destructive evaluation. SPIE, San Diego
20. El-Etriby AE et al A Multi-scale based model for composite materials with embedded PZT filaments for energy harvesting. In: Karaman I, Arróyave R, Masad E (eds) Proceedings of the TMS Middle East—Mediterranean materials congress on energy and infrastructure systems (MEMA 2015). Springer International Publishing, Cham, pp 361–379
21. Hagood IV NW, Bent AA (1999) Composites for structural control. U.S. Patent 5,869,189, 9 Feb 1999
22. Bent AA, Hagood NW, Rodgers JP (1995) Anisotropic Actuation with piezoelectric fiber composites. Intell Mat Sys Struct 6:338–349
23. Brockmann TH (2009) Theory of adaptive fiber composites: from piezoelectric material behavior to dynamics of rotating structures, Vol 161. Springer Science & Business Media, Germany
24. Choi S-C, Park J-S, Kim J-H (2007) Vibration control of pre-twisted rotating composite thin-walled beams with piezoelectric fiber composites. J Sound Vib 300:176–196
25. Sodano HA, Park G, Inman DJ (2004) An investigation into the performance of macro-fiber. Mech Syst Signal Process 18:683–697
26. Hatem TM, Abdel-Meguid M (2014) A multiscale-based model for composite materials with embedded PZT filaments. In: Proceedings of the 2014 International Conference on Power Systems, Energy, Environment. Interlaken, Switzerland
27. Eshelby JD (1957) The determination of the elastic field of an ellipsoidal inclusion, and related problems. In: Proceedings of the royal society of London A: mathematical, physical and engineering sciences. The Royal Society
28. Mori T, Tanaka K (1973) Average stress in matrix and average elastic energy of materials with misfitting inclusions. Acta Metall 21(5):571–574
29. Dunn M, Taya M (1993) Micromechanics predictions of the effective electroelastic moduli of piezoelectric composites. Int J Solids Struct 30(2):161–175
30. Li JY, Dunn ML (1998) Micromechanics of magnetoelectroelastic composite materials: average fields and effective behavior. J Intell Mater Syst Struct 9(6):404–416
31. Benveniste Y (1994) Exact results concerning the local fields and effective properties in piezoelectric composites. J Eng Mater Technol 116(3):260–267
32. Schulgasser K (1992) Relationships between the effective properties of transversely isotropic piezoelectric composites. J Mech Phys Solids 40(2):473–479
33. Kari S et al (2007) Numerical evaluation of effective material properties of transversely randomly distributed unidirectional piezoelectric fiber composites. J Intell Mater Syst Struct 18(4):361–372
34. Bahei-El-Din Y (2009) Modelling electromechanical coupling in woven composites exhibiting damage. Proc Inst Mech Eng Part G: J Aerospace Eng 223(5):485–495
35. Shalan KM et al (2014) Multiscale model and experimental study of damage in piezoelectric fiber-based composite. In: EWSHM—7th european workshop on structural health monitoring. HAL, Nantes, France
36. Hatem TM, Abuelfoutouh MN, Negm HM (2005) Structural health monitoring of metallic structures using modal properties. J Eng Appl Sci 52(3):609–623
37. Hatem TM, Abulfoutouh NM, Negm HM (2004) Application of genetic algorithms (GA) and neural networks (NN) to health monitoring of composite structures. In: 2nd European workshop on structural health monitoring. Munich, Germany
38. Advanced Cerametrics Inc. Piezoelectric fiber composites. Product Catalogue, January 27 h, 2014. Available from: http://www.advancedcerametrics.com/products/energy-harvesting/piezoelectric-fiber-composites/

39. Farrar CR, Worden K (2013) Structural health monitoring: a machine learning prespective. Wiley, USA
40. IEEE, Standard 176 (1987), Institute of Electrical and Electronics Engineers Standard on Piezoelectricity. ANSI/IEEE
41. Berlincourt D, Krueger HHA, Near C (2013) Technical publications 226: important properties of morgan electro ceramics piezoelectric ceramics (PZT-4, PZT-5A, PZT-5H, PZT-8). Morgan Electro Ceramics

Part IV
Mechanical Behavior of Metal Matrix Composites

Effect of Matrix Properties and Sliding Counterface on the Wear Behavior of Magnesium Alloy Metal Matrix Composites

S. Jayalakshmi, R. Arvind Singh and T. S. Srivatsan

Abstract Sliding wear behavior of two magnesium alloys, namely AM100 and ZC63, along with their saffil alumina short fiber-reinforced composites, produced by the technique of squeeze infiltration, is the focus of this technical manuscript. The wear tests were conducted against two different types of counterface materials, namely: (i) EN24 Steel, and (ii) SiC abrasive discs, using a pin-on-disc tribometer. Results revealed that for both the discs, the magnesium alloy based metal-matrix composites revealed enhanced wear resistance when compared one-on-one with the unreinforced counterpart. For both the chosen composite systems it was observed that against the EN24 steel disc, the wear rate decreases with an increase in fiber volume fraction, while against the SiC abrasive disc, the wear rate revealed an increase after certain volume fraction of the fiber reinforcement in the magnesium alloy metal matrix. Against both types of counter-faces, the magnesium alloy AM100 and its composite counterpart showed lower wear rate when compared one-on-one with the magnesium alloy ZC63 and its composite counterpart. The wear behavior of the chosen magnesium alloys, i.e., AM100 and ZC63 and their composite counterparts is discussed considering the conjoint and mutually interactive influences of nature of the matrix (brittle/ductile), characteristics of the counterface and the role of debris as 'third body' during sliding.

S. Jayalakshmi
Department of Aeronautical Engineering, Kumaraguru College of Technology (KCT), Coimbatore 641 049, Tamil Nadu, India
e-mail: Jayalakshmi.subramanian@gmail.com

R. Arvind Singh
Department of Mechanical Engineering, Kumaraguru College of Technology (KCT), Coimbatore 641 049, Tamil Nadu, India
e-mail: arvindsingh.r@gmail.com

T. S. Srivatsan (✉)
The University of Akron, Akron, OH 44325-2903, USA
e-mail: tsrivatsan@uakron.edu

T. S. Srivatsan et al. (eds.), *Metal-Matrix Composites Innovations, Advances and Applications*, The Minerals, Metals & Materials Series,
https://doi.org/10.1007/978-3-319-72853-7_10

Keywords Magnesium alloys · Alumina fiber · Magnesium composites
Hardness · Wear · Friction · Matrix microstructure · Third body

Ensuring that structures are light in weight has become a critical requirement for a spectrum of applications spanning the industries of aerospace, automotive, space-related, consumer electronics and even sporting goods, primarily because prudent use of light-weight materials does contribute to improving the efficiency specific to energy consumption, fuel consumption and maneuverability [1]. Among the light-weight metals to include titanium, aluminium and magnesium, magnesium (Mg) is by far the lightest, with a density of 1738 kg/m^3 [1, 2]. Pure magnesium and its alloy counterparts have only nominal strength at room temperature (25 °C). An improvement in their properties can be achieved by reinforcing pure magnesium and its alloys with ceramic particles, fibers and even whiskers to form discontinuously-reinforced magnesium alloy-based metal matrix composites (Mg-MMCs). Typically, oxide and carbide ceramics [such as, silicon carbide (SiC), boron carbide (B_4C), and alumina (Al_2O_3)] having high strength and stiffness have been used as potential reinforcements to a soft magnesium alloy metal matrix. The engineered Mg-MMCs have to offer an attractive combination of high strength-to-weight ratio [σ/ρ], high temperature stability, enhanced stiffness, hardness and wear resistance [3].

When two materials undergo relative mechanical motion, friction and wear come into play. Wear can be succinctly defined as the undesirable loss of material that often culminates in catastrophic failures. Through the years both magnesium alloys and their composite counterparts have been proposed for use in applications involving tribological contacts, such as: (i) brakes, (ii) brake drums, and (iii) piston ring-cylinders [1]. Though magnesium-based materials have attractive properties and advantages to offer, their tribological properties have not yet been investigated in detail when compared to materials based on the family of aluminum alloys, both monolithic and composite counterpart. The tribological properties of Mg-alloys and their composites have been investigated for the Mg-Al-Zn system [4–8]. Various wear mechanisms, namely: (i) adhesion, (ii) abrasion, (iii) oxidation, (iv) delamination, (v) thermal softening, and even (vi) melting have been reported for the magnesium-based materials, which manifest depending on the applied normal load and sliding speed [4–8].

In the present work, magnesium alloys AM100 alloy (Magnesium-Aluminum-Manganese) and ZC63 alloy (Magnesium-Zinc-Copper) and their alumina short fiber-reinforced composites, produced using the technique of squeeze casting, were studied for their sliding wear behavior. Wear tests were conducted on two different counterface disc materials. The nature of matrix of the chosen base alloy, fiber volume fraction, third-body effects coupled with characteristics of the chosen counterface disc materials that exert an influence on wear behavior of the two chosen alloys and the engineered composite counterparts is presented and briefly discussed.

Experimental Details

Magnesium alloys AM100 (Mg-9.3 to 10.7Al-0.13Mn) and ZC63 (Mg-5.5 to 6.5 Zn-2.5 to 3.5 Cu and 0.25 to 0.75 Mn) and their saffil alumina short fiber-reinforced composites were produced using the techniques of squeeze casting and squeeze infiltration. The saffil alumina short fiber preforms (provided by Imperial Chemical Industries, Japan) having a diameter of 70 mm and height of 35 mm and three different volume fractions, namely: 15, 20 and 25%, were used. Both the alloys and the composite counterpart were produced using a squeeze pressure of 40 MPa. For both the base alloys and the engineered composites the aging cycle was optimized, and the materials were carefully heat-treated to the T6 condition to achieve peak hardness. The microstructure of both the unreinforced alloys and their composite counterparts were studied using both optical microscopy and scanning electron microscopy (SEM).

The hardness of the unreinforced magnesium alloys and the engineered composites was determined using a Brinell hardness tester having a 5 mm-ball indenter and at load of 500 kgf. The hardness tests were conducted on five specimens for purpose of ensuring repeatability.

The wear experiments were conducted under dry sliding conditions at ambient temperature (25 °C) using a pin-on-disc machine. The sliding speed and the applied normal load were kept constant at 1.5 m/s and 30 N, respectively. Experiments were conducted against two different counterface materials, namely:

(a) Hardened EN24 steel disc (hardness: 52 HR_c, roughness: 0.2 μm), and
(b) Silicon carbide abrasive disc (hardness: 20 GPa, roughness: 18.5 μm).

Cylindrical pins of the two chosen magnesium alloys and their composite counterparts having a diameter of 6 mm were used as the test specimen. All specimens were used in the T6 heat-treated condition. Wear, in terms of loss in height of the chosen test specimen, was measured using a Linear Variable Differential Transformer (LVDT), which had a range of 2000 μm with ±1 μm resolution. Frictional force was measured using a load cell capable of measuring up to 20 N. The experiments were repeated three times, and plotted values of both wear and friction are the averages taken from three trials. The worn surfaces of the test samples and the worn tracks on the counterface discs were observed in both an optical microscope and a scanning electron microscope with the purpose of identifying the operating wear mechanism.

Results and Discussion

Microstructure

The microstructure of the chosen unreinforced magnesium alloys, i.e., AM100 and ZC63 alloys, in the peak hardened condition is shown in Fig. 1. The microstructure of alloy AM100 reveals the presence of eutectic (α-Mg matrix + β-$Mg_{17}Al_{12}$) both at and along the grain boundaries [9]. Upon aging, the hard and brittle β-$Mg_{17}Al_{12}$ precipitates form well within the eutectic, and in the process producing a network along the grain boundaries (Fig. 1a). The microstructure of alloy Mg-Zn-Cu (Fig. 1b) consists of α-Mg matrix and the Mg $(Zn, Cu)_2$ eutectic compound (Laves phase [10, 11]) present along the grain boundaries, which spheriodized upon aging (see inset in Fig. 1b). The $MgZn_2$ precipitates formed during aging were observed to be distributed homogenously through the matrix (not seen here, as the $MgZn_2$ precipitates are very fine and can only be observed at high magnification in a transmission electron microscope (TEM) [12]).

Representative microstructures of the AM100 magnesium alloy and ZC63 magnesium reinforced with alumina fibers ($V_f = 25$ vol% and $V_f = 15$ vol%, respectively) are shown in Fig. 2. The fibers were uniformly distributed through the magnesium alloy metal matrix. Unlike the AM100 composite, the ZC63 composite did reveal a sizeable number of fibers to be protruding over the polished surface. This is ascribed to be due to the relatively soft and ductile nature of the ZC63 matrix, which on removal due to polishing exposes the ends of the fibers on the polished surface (Fig. 2b).

Hardness

Hardness of the magnesium alloy AM100 and magnesium alloy ZC63 and the engineered composites as a function of fiber volume fraction ($\%V_f$) is shown in

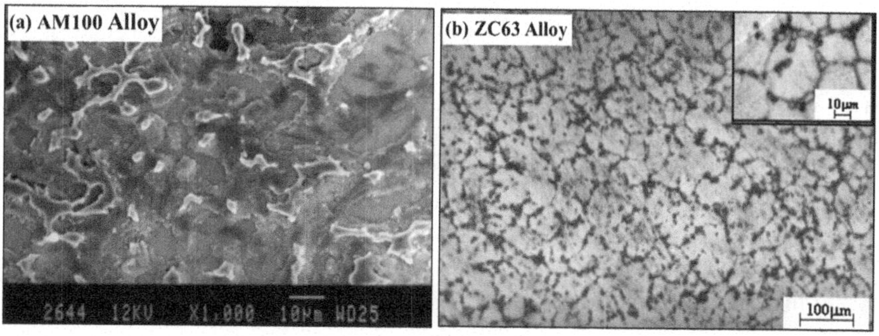

Fig. 1 Microstructure of unreinforced (**a**) AM100 alloy, and (**b**) ZC63 alloy, respectively. The inset shows spherodization of the Mg $(Zn, Cu)_2$ eutectic upon aging

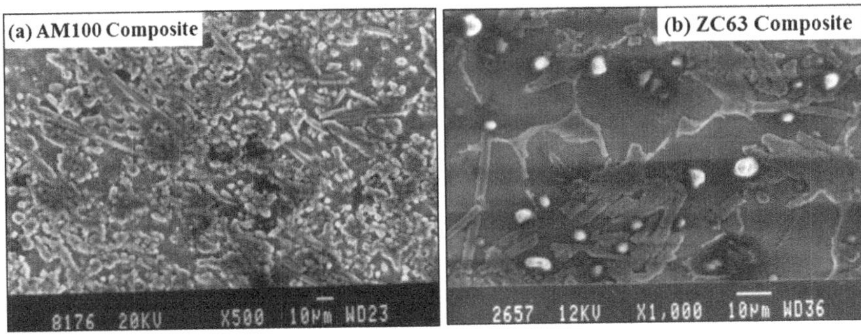

Fig. 2 Distribution of alumina short fiber reinforcement in (**a**) AM100 (25% V_f) matrix, and (**b**) ZC63 (15% V_f) matrix

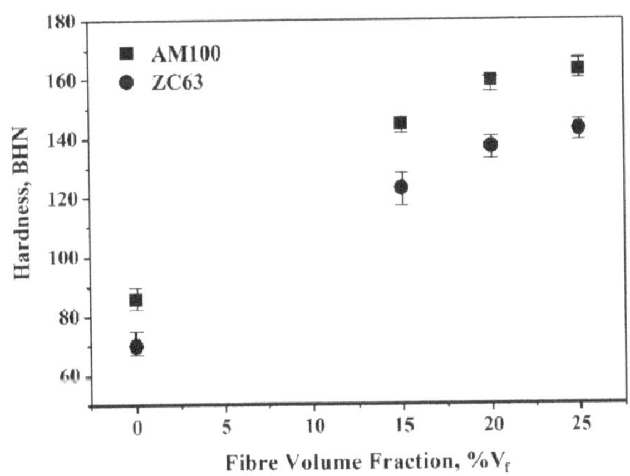

Fig. 3 Hardness of AM100 and ZC63 alloys and their composites with an increase in volume fraction of the saffil alumina fiber reinforcement

Fig. 3. In both the alloy systems, the engineered composites revealed higher hardness than the unreinforced alloys due to the high inherent hardness of the reinforcing fiber. Hardness of the engineered composite increased with an increase in fiber volume fraction, as the amount of the hard reinforcement phase in a soft metal matrix increases. The presence of hard ceramic reinforcement increases the dislocation density at the fiber/matrix interface, which arises as a direct consequence of residual stresses caused by thermal expansion mismatch between the reinforcing fibers and the soft alloy matrix [13]. Upon aging, interface (with high activation energy) acts as a potential nucleation site for precipitation to occur [14], thereby contributing in a positive way to increasing the overall hardness.

It is also observed that the AM100 alloy is harder than the ZC63 alloy, and similarly the AM100 composites were noticeably harder than the ZC63 composites. The high hardness of AM100 alloy can be attributed to the high inherent hardness of the $Mg_{17}Al_{12}$ precipitates (280 Hv [15]). For the engineered AM100 composites, the high hardness can be ascribed to the formation and presence of $Mg_{17}Al_{12}$ precipitates at the fiber/matrix interfaces during aging.

Wear and Friction

Sliding Against EN24 Steel Counterface Disc

The volumetric wear loss per unit sliding distance (m^3/m) of magnesium alloy AM100 and magnesium alloy ZC63 and their composite counterparts when slid against EN24 steel counterface disc is shown in Fig. 4. The influence of fiber reinforcement on wear behavior can be identified by comparing the wear rates of the reinforced alloys with the unreinforced counterpart. It can be seen from the Fig. 4 that both the composites have wear rates that are significantly lower than those of the unreinforced counterpart. This is due to the fact that wear is inversely proportional to hardness (Archard's wear equation) [16]. With an increase in fiber volume fraction in the magnesium alloy metal matrix, the hardness increases (Fig. 3) causing further reduction in wear rate. It is observed that the wear rates of magnesium alloy AM100 and its composite counterpart are almost one-half of those of magnesium alloy ZC63 and its composite counterpart. The lower wear rates of the AM100 alloy can be attributed to their high inherent hardness caused by the presence of hard $Mg_{17}Al_{12}$ precipitates.

Representative worn surfaces of the unreinforced alloy AM100 and its composite counterpart containing 20% V_f fibers is shown in Fig. 5a, b. The alloy (Fig. 5a) reveals micro-ploughing and adhesion, while the composite (Fig. 5b) reveals evidence of delamination. In Fig. 5b is shown a delaminated region containing powdered wear debris due to repeated sliding.

Representative worn surfaces of magnesium alloy ZC63 and its composites are shown in Fig. 6a, b. Material removal (in process) due to the adhesion in alloy ZC63 is shown in Fig. 6a, while delamination wear in the composite is shown in Fig. 6b. Delamination occurs due to the formation and presence of subsurface cracks and their ensuing propagation through the microstructure [17]. The cracks originate in the composites due to fragmentation of the fiber during sliding.

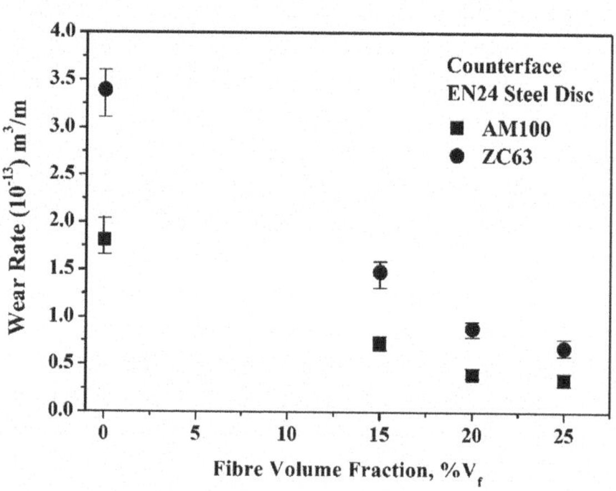

Fig. 4 Wear rate as a function of fiber volume fraction for AM100 and ZC63 alloys and their alumina short fiber-reinforced composites against EN24 steel counterface disc

Fig. 5 Scanning electron showing representative worn surfaces of: (**a**) AM100 alloy and (**b**) AM100 composite (20% V_f), When slid against EN24 steel counterface

Fig. 6 Scanning electron micrographs showing the representative worn surfaces of (**a**) ZC63 alloy, and (**b**) ZC63 composite (20% V_f), when slid against EN24 steel counterface

Once the cracks are initiated, further sliding causes the cracks to extend and propagate through the microstructure, and in the process eventually link and/or coalesce with the neighboring cracks [18]. These cracks propagate parallel to the sliding direction and as they reach the surface, material in the form of thin sheets tend to delaminate [19].

Sliding Against SiC Abrasive Counterface Disc

The volumetric wear loss per unit sliding distance (m^3/m) of both the alloys and the engineered composites when slid against SiC abrasive disc is shown in Fig. 7. Similar to the trend observed when slid against an EN24 steel disc, magnesium alloy AM100 revealed observably higher wear resistance than magnesium alloy ZC63 when slid against a SiC counterface. This can essentially be attributed to the high inherent hardness of the magnesium alloy AM100 (Fig. 3).

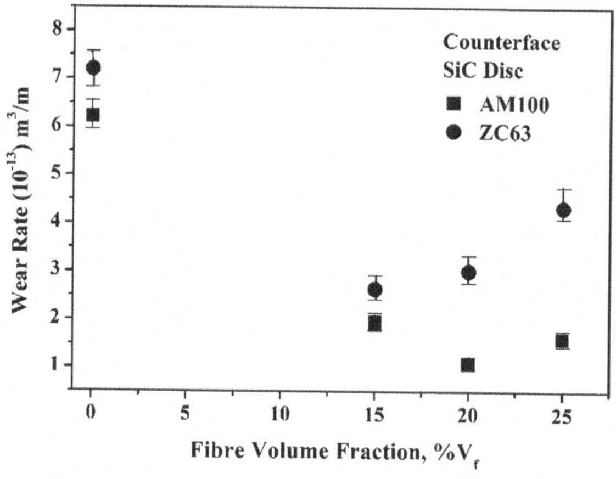

Fig. 7 Wear rate as a function of fiber volume fraction in AM100 and ZC63 composites against SiC abrasive counterface disc

Further, it is observed that both the composites of the two chosen magnesium alloys, i.e., AM100 and ZC63, have wear rates that are lower than those of their unreinforced counterpart. The engineered composites exhibit higher wear resistance due to their higher hardness when compared to the base alloy (Fig. 3).

As was observed earlier, when slid against the EN24 steel disc, the wear rates of the composites of the two chosen magnesium alloys tend to decrease with an increase in volume fraction of the fiber reinforcement (Fig. 4). Interestingly, such a linear trend does not exist when the composites are slid against the SiC disc. Against the SiC abrasive disc, the wear rates tend to initially decrease. With an increase in volume fraction of the fiber reinforcement, an increase in wear rate was observed for the magnesium alloy AM100 composite for 25% V_f and for the magnesium alloy ZC63 composite for 20% V_f. Such an observation has also been reported by other researchers [20–22]. The reason for the relative increase in wear rates for higher volume fraction of the fiber reinforcement in the magnesium alloy metal matrix is discussed in the following sections.

Worn surfaces of the AM100 alloy and its composite counterpart containing 20 vol% of the fiber reinforcement when slid against SiC abrasive counterface disc are shown in Fig. 8a, b. Compared to the EN24 disc, the SiC abrasive disc has higher roughness and hardness (see Sect. 2). Given this fact, the material removal in the AM100 alloy occurs predominantly by ploughing as is evident from Fig. 8a (evidence of adhesion was absent). Composites of the AM100 alloy experience wear by delamination as seen in Fig. 8b, by the mechanism that was explained earlier. The worn surfaces of magnesium alloy ZC63 and its composite when slid against the SiC abrasive disc is shown in Fig. 9. Wear experienced by the magnesium alloy ZC63 occurs due to extensive ploughing by the hard grits on the counterface of the SiC disc (Fig. 9a), since the matrix is essentially ductile [23]. Delamination leads to material removal in the composite (Fig. 9b).

The wear tracks on SiC abrasive counterface due to the sliding of alloy ZC63 (marked ‘A’) and alloy AM100 (marked ‘B’) composites are shown in Fig. 10.

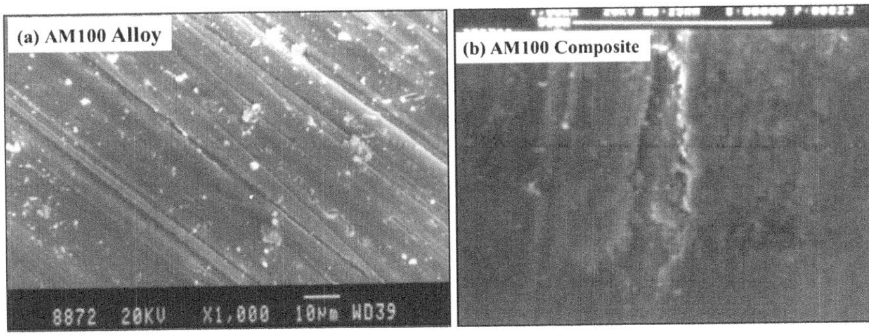

Fig. 8 Worn surfaces of: (**a**) Magnesium alloy AM100, and (**b**) Magnesium alloy AM100 composite (containing 20 vol% of fibers) when slid against a SiC disc

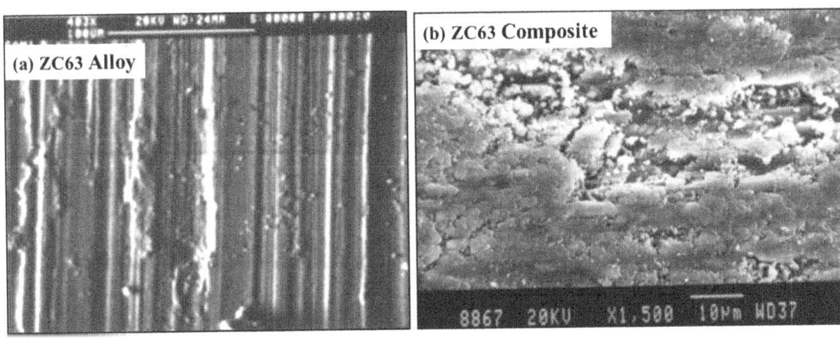

Fig. 9 Worn surfaces of (**a**) Magnesium alloy ZC63, and (**b**) Magnesium alloy ZC63 composite (containing 20 vol% of reinforcing fibers) when slid against the SiC disc

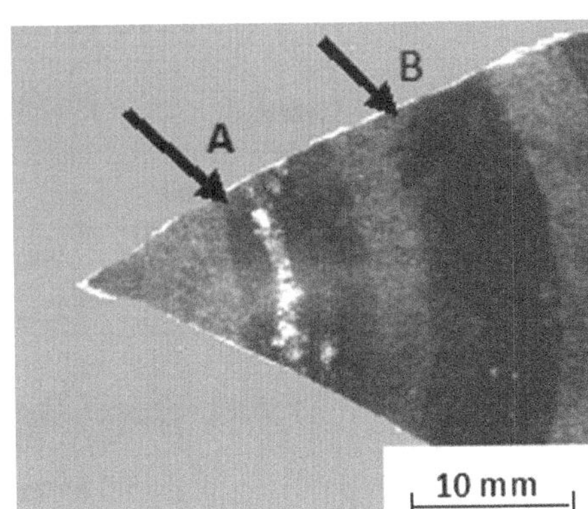

Fig. 10 Photograph showing wear tracks of the 20% V_f composites of: (**A**) Magnesium alloy ZC63, and (**B**) Magnesium alloy AM100

Smearing of the pin material is seen on the wear track produced by the magnesium alloy ZC63 based composite (denoted as white region on track ‘A’), which was absent for the case of the magnesium alloy AM100 based composite.

The role of wear debris in the wear process of composites when slid against the SiC counterface is shown in Fig. 11. As seen in Fig. 11a for composites of magnesium alloy AM100, wear debris, which contains brittle matrix material and broken fibers are loosely trapped in the grits of the SiC counterface. During sliding, the free debris abrades the pin surface and rolls over along the sliding direction, and eventually gets released from the surface. In contrast, during sliding of the composite of magnesium alloy ZC63 (Fig. 11b), since the matrix is essentially ductile [23], the wear debris gets embedded between the abrasive SiC grits and gradually gets compacted due to repeated sliding. During this process, the broken reinforcing fibers, which are hard, brittle and experience elastic deformation get trapped in the compacted material on the wear track (shown as “white” region on the track ‘A’, Fig. 10). This ‘mechanically locked’ third body along with the hard-abrasive grits severely counter-abrades pins of the composite. Therefore, an increase in wear was observed with an increase in volume fraction of the reinforcing fibers against the

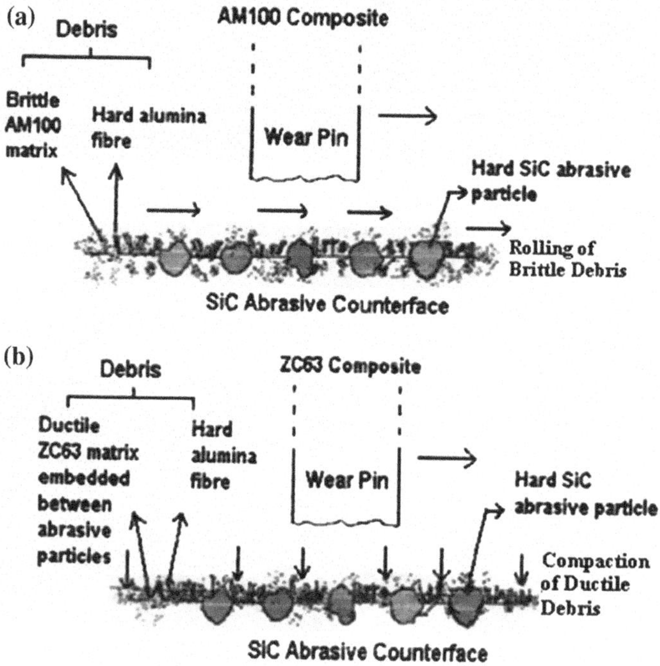

Fig. 11 Schematic diagrams showing the role of wear debris in the wear process of composites against SiC counterface. **a** Wear debris of AM100 composite that are loosely trapped undergo rolling during sliding, and **b** Wear debris of ZC63 composite compacted at the rough surface of SiC abrasive disc trap hard alumina fibers\which severely counter-abrade the composite pin during sliding

SiC abrasive counterface disc (Fig. 7). The above discussion highlights the influence of the following factors:

(a) The nature of the matrix (ductile/brittle).
(b) Volume fraction of reinforcement.
(c) Wear debris as 'third-body' and
(d) Surface characteristics of the counterface discs,

in determining the wear behavior of the two chosen magnesium alloys AM100 and ZC63 and their composite counterparts.

Coefficient of Friction

The coefficient of friction (μ) as a function of volume fraction of fiber reinforcement for the two chosen magnesium alloys and their composite counterparts when slid against the EN24 steel disc and the SiC abrasive disc is shown in Fig. 12.

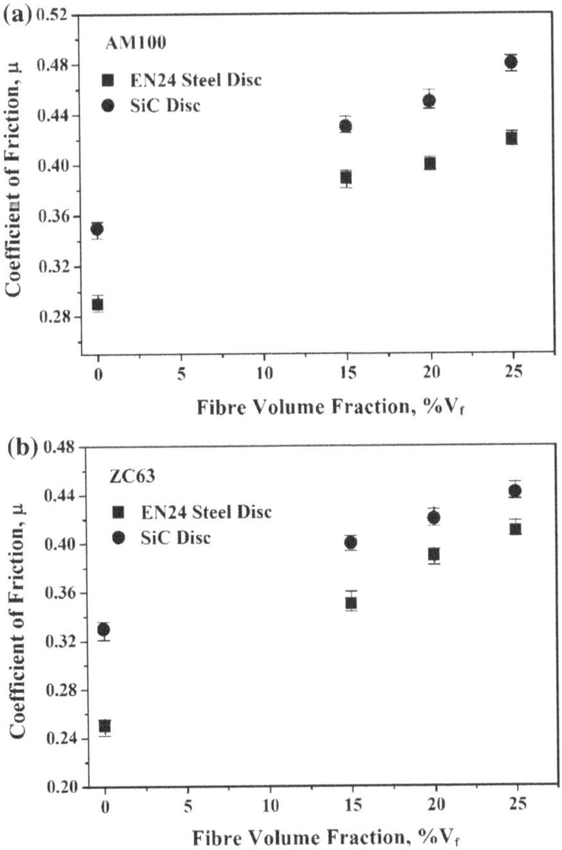

Fig. 12 Coefficient of friction as a function of fiber volume fraction. **a** Magnesium alloy AM100 and composites and **b** Magnesium alloy ZC63 and composites, when slid against the EN24 steel disc and a SiC abrasive counterface disc

Against both the counterface, the engineered composites exhibit a higher value μ than the unreinforced alloy due to the conjoint action of the abrasive action induced by the hard-ceramic reinforcement and the occurrence of counter-abrasion by the wear debris. For the case of the SiC counterface, abrasion of the pin material by the hard grits also contributes to increasing the value of coefficient of friction (μ). Hence, it is observed that for both the base alloys and their composites, the values of coefficient of friction (μ) are higher when slid against SiC than when slid against EN24 steel disc. For the composites, the value of coefficient of friction (μ) increases with an increase in volume fraction of the ceramic reinforcement. To note, the composites show lower wear rates than the unreinforced matrix alloy, whereas their values for the coefficient of friction (μ) are higher than those of the unreinforced counterpart, i.e., the base alloy. This 'low wear—high friction' combination of tribological characteristic is a preferred requirement for materials chosen for use in brakes.

Conclusions

The wear behavior of the chosen magnesium alloys AM100 and ZC63 and their saffil alumina short fiber-reinforced composites when slid against a EN24 steel disc and a SiC abrasive disc were investigated. The following are the key conclusions that can be drawn from this study.

(1) The wear resistance of the composites against both the EN24 steel and SiC abrasive counterface discs is higher when compared to the base alloys. This is essentially due to their high hardness caused by the presence of the hard, brittle and essentially elastically deforming ceramic fibers.
(2) Against the EN24 steel disc, composites of the two magnesium alloys AM100 and ZC63 revealed an observable decrease in the wear rate with an increase in fiber volume fraction. This is essentially because of an increase in hardness arising because of an increase in the amount of the hard reinforcement phase. When slid against the EN24 steel, the AM100 magnesium alloy revealed higher wear resistance when compared one-on-one to the magnesium alloy ZC63 due to their high inherent hardness caused by the presence of hard $Mg_{17}Al_{12}$ precipitates.
(3) Against the SiC abrasive disc, the wear rates of composites based on the two magnesium alloys AM100 and ZC63 increased following a certain volume fraction of fibers as a direct consequence of counter abrasion arising from the third body (debris).
(4) During wear of the engineered magnesium composites against the SiC abrasive disc, the role of the wear debris as the 'third body' in influencing wear depends on the nature of the magnesium alloy chosen. The loosely trapped debris (hard $Mg_{17}Al_{12}$ + ceramic fibers) in composites of magnesium alloy AM100 tend to counter-abrade the pins and eventually roll over to get released from the surface

during repeated sliding. For the case of the composite of magnesium alloy ZC63, the ductile matrix gets compacted in the grits of the SiC abrasive disc, which traps the hard-ceramic fibers causing counter-abrasion during repeated sliding.

(5) Worn surfaces of the chosen magnesium alloys AM100 and ZC63 revealed wear to occur by adhesion and ploughing. The composite counterparts of the two chosen alloys experienced wear by delamination.
(6) Wear of AM100 and ZC63 alloys and their composites was influenced by: (a) the nature of the matrix (ductile/brittle), (b) volume fraction of reinforcement, (c) wear debris as 'third-body' and (d) surface characteristics of the counterface discs.
(7) The coefficient of friction of the composites against the two chosen counterface was higher than for the base alloy due to the occurrence of abrasion of the disc by the hard-ceramic fiber reinforcements, which increased with an increase in volume fraction of the reinforcement. The coefficient of friction of both the alloy systems was higher against the SiC abrasive disc when compared to that against the EN24 steel disc. This can be attributed to the abrasion of the pins by the hard grits of the counterface disc.

References

1. Ceschini L, Dahle A, Gupta M, Jarfors AA, Jayalakshmi S, Morri A, Rotundo F, Toschi S, Arvind Singh R (2016) Aluminium and magnesium metal matrix nanocomposites. Springer Nature, Singapore
2. Kainer K (2006) Metal matrix composites: custom-made materials for automotive and aerospace engineering. Wiley, UK
3. Gupta MM, Sharon NML (2011) Magnesium, magnesium alloys, and magnesium composites. Wiley, New Jersey
4. Taltavull C, Torres B, Lopez AJ, Rams J (2013) Dry sliding wear behavior of AM60B magnesium alloy. Wear 301:615–625
5. Das S, Morales AT, Alpas AT (2010) Microstructural evolution during high temperature sliding wear of Mg–3% Al–1% Zn (AZ31) alloy. Wear 268:94–103
6. Blau PJ, Walukas M (2000) Sliding friction and wear of magnesium alloy AZ91D produced by two different methods. Tribol Int 33:573–579
7. Mehta DS, Masood SH, Song WQ (2004) Investigation of wear properties of magnesium and aluminum alloys for automotive applications. J Mater Process Technol 155–156:1526–1531
8. Lim CYH, Lim SC, Gupta M (2003) Wear behavior of SiCp-reinforced magnesium matrix composites. Wear 255:629–637
9. Jayalakshmi S, Kailas SV, Seshan S (2002) Tensile behavior of AM100 magnesium alloy and its Al_2O_3 fiber reinforced composites. Composites-A 33:135–140
10. Nunez-Lopez CA, Skeldon PS, Thompson GE, Lyon P, Karimzadeh H, Wilks TE (1995) The corrosion behavior of Mg alloy ZC71/SiC metal matrix composite. Corros Sci 37(5):689–708
11. Komura Y, Mitarai M, Nakatani Iba H, Shimuzu T (1970) Structural changes in the alloy systems of Mg-Zn-Cu and Mg-Zn-Ag related to the Friauf-Laves phases. Acta Crystallogr A B26:666–668
12. Yang W, Weatherly GC, Mccomb DW, Llyod DJ (1997) The structure of SiC-reinforced Mg casting alloys. J Microsc 185:292–302

13. Arsenault RJ, Shi N (1986) Dislocation generation due to differences between the coefficients of thermal expansion. Mater Science Eng 81:175–187
14. Arsenault RJ, Fischer RM (1989) Microstructure of fiber and particulate SiC in 6061 Al composite. Scr Metall 17:67–71
15. Alderman M, Manuel MV, Hort N, Neelameggham NR (eds) (2014) Magnesium technology 2014. The Minerals, Metals & Materials Society. Warrendale, USA
16. Archard JF (1953) Contact and rubbing of flat surfaces. J Appl Phys 24:981–988
17. Suh NP (1986) Tribophysics. Prentice Hall Inc., New Jersey
18. Alahelisten A, Bergman F, Olsson M, Hogmark S (1993) On the wear of aluminum and magnesium metal matrix composites. Wear 165:221–226
19. Hosking FM, Pertillo SV, Wanderlin R, Mehrabian R (1982) Composites of aluminum alloys: fabrication and wear behavior. J Mater Sci 17:477–498
20. Hughes D (1988) Textron unit makes reinforced Ti, Al parts. Aviation Week Space Technology, USA
21. Westengen H, Albright DL, Nygard A (1990) Development of cast Mg-matrix composites. SAE Tech. Paper 900534:606–612
22. Hutchings IM (1994) Tribological properties of metal matrix composites. Mater Sci Technol 10:513–517
23. Jayalakshmi S, Kailas SV, Seshan S, Fleury E (2006) Properties of squeeze cast Mg-6Zn-3Cu alloy and its saffil alumina short fiber reinforced composites. J Mater Sci 41:3743–3752

Synthesis and Microstructural Development of Particulate Reinforced Metal-Matrix Composites Using the Technique of Spray Atomization and Deposition

T. S. Srivatsan, Yaojun Lin, Fei Chen, K. Manigandan and Enrique J. Lavernia

Abstract In this paper, the synthesis of discontinuously-reinforced metal matrix composites using the technique of spray atomization and co-deposition is presented and briefly discussed. This technical manuscript provides a lucid overview of the salient features of the technique of spray atomization and deposition. An attempt is made to highlight the key attributes specific to the processing technique with specific reference to pre-mixed metal matrix composites. The key factors and viable mechanisms governing incorporation of the reinforcing phase during co-deposition is presented and adequately discussed. The microstructure of few composites synthesized by this technique and the key and observable features are briefly highlighted considering the conjoint and mutually interactive influences of processing variables and constituents of the metal-matrix composite.

Keywords Spray atomization and deposition · Metal-matrix composites
Discontinuously-reinforced · Processing techniques · Processing variables

T. S. Srivatsan (✉)
The University of Akron, Akron, OH 44325-2903, USA
e-mail: tsrivatsan@uakron.edu

Y. Lin · F. Chen · E. J. Lavernia
Chemical Engineering and Materials Science, University of California, Irvine, 509 Aldrich Hall, Irvine, CA 92697-1000, USA

K. Manigandan
Department of Mechanical Engineering, The University of Akron, Akron, OH 44325-2903, USA

T. S. Srivatsan et al. (eds.), *Metal-Matrix Composites Innovations, Advances and Applications*, The Minerals, Metals & Materials Series,
https://doi.org/10.1007/978-3-319-72853-7_11

Introduction

The continuing need for lightweight and high-performance materials to satisfy both the numerous demands and burgeoning needs put forth by the industries spanning air transportation, ground transportation, sporting goods and commercial products has in recent years provided the much-needed impetus for a plethora of research studies. This has, since the early 1990s, eventually led to the development and emergence of new and improved materials resulting from novel and innovative combinations of the conventional materials. During the time spanning the last four decades, i.e., since the early 1970s, the preponderance of research efforts has focused on both alloy design and use of novel and/or innovative processing techniques to synthesize high-performance hybrid materials, termed as composites, as potentially viable and economically affordable alternatives to the traditional and newer generation of unreinforced metal alloys [1–10].

A sizeable number of these composite materials are metallic matrices reinforced with high strength, high modulus and often brittle second phases, which are either continuous in the form of fibers or discontinuous in the form of whisker, short-fiber, platelet and particulate reinforcements embedded in a ductile metal matrix. Incorporation of hard, brittle and elastically deforming ceramic reinforcements in a soft, ductile and essentially plastically deforming metal matrix has been shown in far more than few cases to offer improvements in elastic modulus, wear resistance, tensile strength, structural efficiency, reliability and in certain instances fatigue resistance while concurrently controlling the physical properties, such as: (i) density, and (ii) coefficient of thermal expansion [11–15]. An appropriate combination of these properties has made possible improved mechanical properties of the engineered composite when compared one-on-one with the unreinforced metal matrix [11–20]. In fact, enhanced interest in the selection and use of metal-matrix composites for the industries spanning aerospace, automotive and other ground transportation industries, and the rapidly growing industry of commercial products has gained increased importance because of an availability of relatively inexpensive reinforcements coupled with the concurrent development of various processing routes, which often result in reproducible microstructures and properties [21–23].

A composite, also referred to as a hybrid material by individuals populating both the scientific and technical communities, can be best defined as a heterogeneous structural material consisting of two or more distinct conventional materials, which are either mechanically or metallurgically bonded together to achieve a definite goal for a specific purpose [18–20]. The primary purpose behind the development and commercialization of composite materials was the following:

(i) To synergize the desirable properties of the constituents, that is (a) the reinforcement or the second-phase, and (b) the bulk or parent material called the matrix (the primary phase).
(ii) To suppress the shortcomings of each of the constituents.

This combination can result in a newly synthesized material having a unique combination of properties for a spectrum of structural and even non-structural applications [19–21]. Interest in the use of reinforcements in a continuous matrix phase, dating back to the early 1960s, was essentially due to one or more limitations imposed by the conventional ingot metallurgy (IM) processing technique coupled with a grueling need for a newer generation of materials to meet the ever-increasing demands put forth by the industries spanning space, ground transportation and commercial products, which essentially required a radical new approach to material synthesis and preparation [20, 24–26].

The reinforcement can be either a continuous phase or a discontinuous phase. Continuous reinforcements are typically fibers and occasionally laminates. The discontinuous reinforcements include the following: (i) particulates, (ii) platelets, (iii) chopped fibers, and (iv) whiskers. Typical reinforcement geometries are shown in Fig. 1 [17, 18, 20]. The discontinuous reinforcements are advantageous in that the composites are relatively easy to synthesize using the techniques of: (a) powder metallurgy (PM), (b) ingot metallurgy (IM), and (c) mechanical alloying (MA). Further, the discontinuously-reinforced metal matrices are attractive since they exhibit near isotropic properties when compared one-on-one with the continuously-reinforced counterpart [26–31]. The primary disadvantage of all metal-matrix composites, i.e., both continuously-reinforced and discontinuously-reinforced, is that they suffer from low ductility, inadequate fracture toughness, inferior fatigue crack growth and fracture resistance when compared with the unreinforced monolithic alloy [11, 32–37].

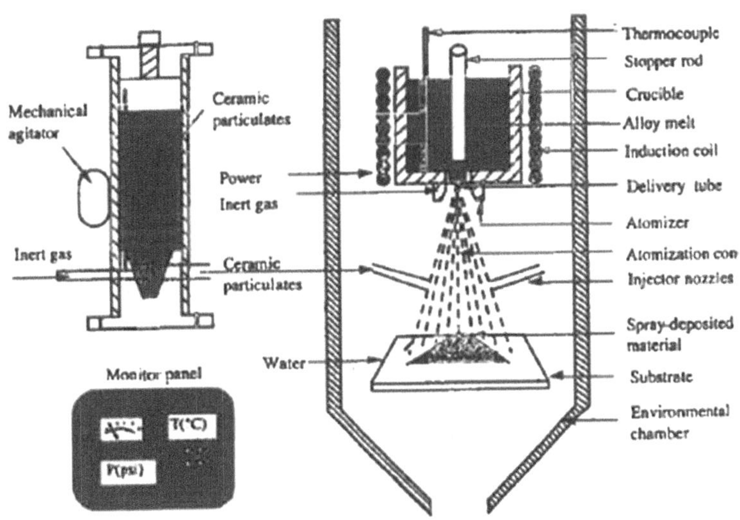

Fig. 1 Schematic diagram showing spray atomization and deposition processing of particulate reinforced metal matrix composites

In more recent years, the discontinuously-reinforced metal matrix composites have engendered considerable scientific and technological interest primarily because of an availability of various types of reinforcements at competitive costs coupled with the successful development of manufacturing processes to produce metal-matrix composites having reproducible microstructures and properties. Besides, standard or near-standard metalworking methods, such as: (a) casting, (b) forging, (c) rolling, and (d) extrusion, have also been successfully used for the synthesis of these materials. Additionally, use of discontinuous reinforcements minimizes many of the problems associated with the fabrication of continuously-reinforced MMCs, such as: (i) fiber breakage, (ii) microstructural non-uniformity, (iii) fiber-to-fiber contact, and (iv) interfacial reactions. However, the discontinuously-reinforced composite materials are not homogeneous and overall material properties are sensitive to the following: (a) properties of the constituents, (b) interfacial properties, and (c) geometric shape of the reinforcement.

An examination of the published literature reveals that a sizeable number of research efforts have been made on the development of novel processing techniques to optimize both the structure and properties of particulate-reinforced metal matrix composites. Of the techniques that have been developed and successfully implemented, spray atomization and deposition processing offers a unique opportunity to synergize the benefits associated with fine particulate technology, that is:

(i) Refinements in intrinsic microstructural features,
(ii) Modifications in alloy chemistry,
(iii) In situ processing, and
(iv) Near-net shape manufacturing.

The technique of spray atomization and deposition processing essentially involves mixing of the discontinuous reinforcements with the matrix material under non-equilibrium conditions. As a result, this processing technique offers an opportunity for enhancing the properties of existing alloy systems while concurrently developing novel alloy compositions.

The primary objective of this paper is to highlight use of the technique of spray atomization and deposition processing for the synthesis of discontinuously-reinforced metal-matrix composites. In particular, emphasis is placed on the influence of spray processing technique on microstructural evolution of metal-matrix composites produced by this technique. The salient features of this technique are highlighted and discussed using appropriate examples.

The Technique of Spray Atomization and Deposition Processing

The technique of spray atomization and deposition processing of discontinuously-reinforced composites is essentially a semi-liquid synthesis methodology in which the chosen reinforcement and the metal matrix are mixed, while the matrix is in a semi-liquid state. The processing technique offers certain unique characteristics that differentiate it from the approaches of liquid, solid, and two-phase processing. Most importantly and interestingly, both mixing and consolidation are completed in a single operation and in the process aids in reducing the number of processing steps. Also, the technique has been used to produce preforms with the as-spray deposited density that exceeds 97% of the theoretical density. This makes the spray deposited metal-matrix composites (MMCs) receptive to densification with limited post-deposition thermomechanical processing. Besides, the technique can also be used for the near-net shape forming of composites in geometrical configurations of billets, plates, tubes, discs, etc. Further, in direct comparison with the liquid phase processes and rheocasting the contact time between the metal matrix and the reinforcement(s) is considerably reduced.

Sustained research and development efforts specific to the technique of spray deposition for the manufacturing of discontinuously-reinforced metal matrix composite materials has during the past three decades, since the early 1990s, engendered both scientific and technological interest. The preponderance of research activity has focused on examining and rationalizing salient aspects of the processing technique. These include the following:

(a) Methods for incorporation of the reinforcing phase or phases, if more than one.
(b) An interaction behavior between the metal matrix and the reinforcing phase.
(c) Microstructural evolution.

Besides, these processing techniques there are a few other processing techniques that have been developed for the synthesis of discontinuously-reinforced metal-based composite materials. These essentially include the following:

(i) Thermal spray deposition, and
(ii) Plasma spray deposition.

Despite intrinsic differences in the methods used for the generation of semi-liquid/semi-solid droplets, the deposition and consolidation stages involved in these techniques are essentially similar.

Early studies on spray deposition processing of metal-matrix composites was initially reported in the late 1980s and early 1990s and essentially involved the simultaneous spraying of both the metal and ceramic particles, a process referred to as co-deposition [38–47]. These early studies used pure aluminum and commercial aluminum alloys as the matrix material. Since then, a wide variety of matrix-reinforcement combinations have been successfully synthesized using the technique of spray deposition. The matrix materials that have been used include the following:

(A) Aluminum-base alloys, such as: (i) Al–Cu [48–51], (ii) Al–Fe [52], (iii) Al–Li [53–55], (iv) Al–Mg–Si [56–58], (v) Al–Si [59–62], (vi) Al–Ti [63], and (vii) Al–Zn–Mg–Cu [64, 65];
(B) Copper alloys [66–68];
(C) Iron-based alloys [69, 70];

(D) Titanium-based alloys [71, 72]; and even
(E) Intermetallic compounds, such as (i) CoSi [73], (ii) $MoSi_2$ [74, 75] and (iii) Ni_3Al [76–78].

The reinforcements frequently tried and used in spray deposition include the following:

(a) Aluminum oxide (Al_2O_3),
(b) Graphite,
(c) Silicon carbide (SiC),
(d) Titanium diboride (TiB_2),
(e) Titanium carbide (TiC), and
(f) Steel in both powder form and fiber form.

Through the years, several spray deposition processing techniques have been developed for the synthesis of metal-based composite materials. These essentially include the following:

(i) Spray atomization and co-deposition [47, 48, 51, 53–55],
(ii) Spray atomization of pre-mixed MMCs [49, 50, 62],
(iii) Modified Osprey method [40–42],
(iv) Reactive spray deposition [76, 79],
(v) Plasma spray deposition [71, 72, 74, 75], and
(vi) Spray deposition of continuous fiber-reinforced intermetallic matrix composites [71, 72, 78, 80].

Spray Atomization and Co-deposition

This is the first and most frequently used spraying technique that has been applied for the synthesis of composite materials. The term co-deposition is used to describe simultaneous deposition of both the matrix and the reinforcing phases. Experimentally, entraining the reinforcements within a flowing gas, and aiming the gas-reinforcement mixture at a fine dispersion of atomized droplets accomplish co-deposition. In principle, this method is suitable for the synthesis of discontinuously-reinforced composites. An inspection of the published scientific literature reveals that co-deposition was initially used for the fabrication of discontinuous particle-reinforced metal matrix composites (DRMMCs) essentially due to an ease of availability of the reinforcements coupled with low costs associated with this technique. A schematic of the equipment used for the spray atomization and co-deposition of composites is shown in Fig. 1. The technique essentially involves three consecutive steps. These are the following:

(a) Atomization,
(b) Ceramic particle co-injection, and
(c) Deposition.

During spray deposition, the matrix material is atomized into a fine dispersion of droplets using a high velocity gas. While atomization is accomplished in the same manner as that used for unreinforced metals and their alloy counterparts it is sometimes both essential and desirable to modify the relevant process parameters, i.e., (i) gas pressure, and (ii) gas-to-metal flow rate ratio. A schematic of the spray atomization and co-deposition technique is shown in Fig. 2 [81]. Following atomization, the droplets level towards the deposition surface under the combined influence of forces of fluid drag and gravity. Prior to arrival and deposition of the droplets on the deposition surface, ceramic particulates are injected into the atomized spray using one or more gas injectors. The injection allows the ceramic particulates to interact with the atomized droplets. Finally, the spray of droplets interspersed with ceramic particles is collected on a substrate or shaped container. The distance between the atomizer and the injector position is defined as the ***injection distance*** and the distance between the atomizer and the deposition surface is defined as the ***deposition distance***. It is also possible to direct flow of the reinforcing particles directly to the deposition surface. This then makes the ***injection distance*** identical to the ***deposition distance***.

To understand the influence of droplet solidification during ceramic co-injection and deposition, Gupta et al. [55] calculated the solidification history of an average size droplet as a function of flight distance using the basis of a simplified enthalpy

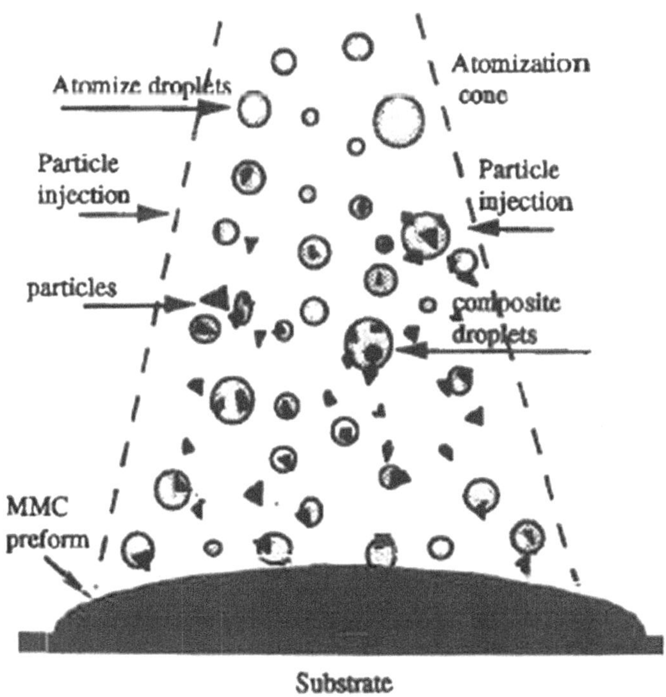

Fig. 2 A schematic illustrating the spray atomization and co-deposition process [58]

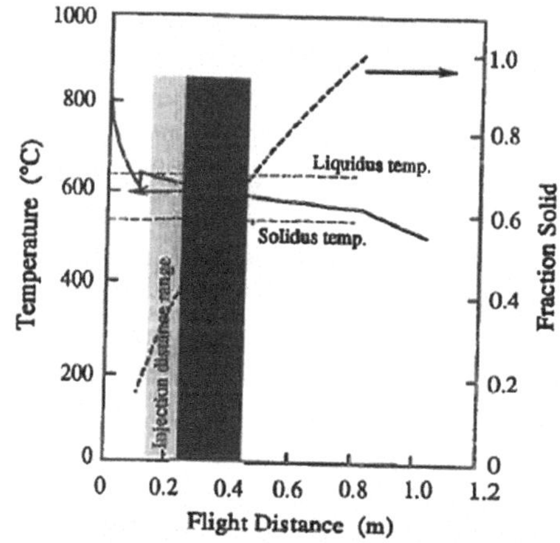

Fig. 3 Calculated temperature and fraction of solid in atomized aluminum-lithium droplets as a function of flight distance [55]

model. The calculated fraction of solid phase and droplet temperature as a function of fight distance for the case of aluminum-lithium droplets, with an average size of 84 μm, is shown in Fig. 3 [55]. The corresponding injection and deposition distances are also shown in this figure. The droplets are in the semi-liquid state at both the ***injection distance*** and ***deposition distance***. Typically, the droplets contain less than 40 vol% of the solid phases at the ***injection distance*** and more than 70 vol% of solid phase at the ***deposition distance*** [55]. The selection of ***deposition distances*** for the discontinuously-reinforced composite is based on the same criteria used for the spray deposition processing of the monolithic alloy, i.e., to produce a high-density spray deposited preform having a refined microstructure. Proper selection of the ***injection distance*** is based on the following considerations:

(1) The selected distance must facilitate proper dispersion of the ceramic particulates to achieve the desired reinforcement volume fraction and distribution. For example, selection of an ***injection distance*** that is too close to the atomizer can lead to freezing of the nozzle. Alternatively, selection of an ***injection distance*** that is close to the ***deposition distance*** will result in low yield of the injected particles.
(2) The ***injection distance*** must be selected to either minimize or avoid interfacial reactions that can occur as the particles come in contact with both the liquid and partially liquid droplets.

In spray atomization and co-injection, the ceramic particulates are accelerated and directed towards the drop spray using gas injectors. Generally, two or more injectors may be used in a single experiment depending on the desired volume fraction of the chosen ceramic reinforcement. Although a variety of designs have been used to entrain the ceramic particulates, using appropriate gas flow control, there are two basic types of injectors that have been used for this purpose, namely:

(a) A fluidized bed injector, and
(b) A coaxial injector.

The injectors are generally located outside of an environmental chamber at a location where gas injection is facilitated directly into the chamber. In Fig. 4 is shown a typical fluidized bed injector [63]. In a fluidized bed injector, the reinforcing ceramic particulates are placed inside of a tube containing a porous material. During injection, a pressurized gas (typically inert) passes through the fluidized bed and entrains the reinforcing ceramic particles. As the gas pressure increases the fluidized bed begins to expand. This is accompanied by a concurrent decrease in bulk density. When a critical pressurizing condition is achieved the drag force that is exerted by the fluid on the reinforcing particulates equals the gravitational force that is holding these particulates together inside the injector container. This condition is referred to as the minimum fluidization condition. Under the minimum fluidization condition, the reinforcing particles are free to move relative to each other in the container. The primary disadvantage associated with fluidized bed injector is a relatively low mass flow rate of the reinforcing particulates. The mass flow rate of the reinforcing particulates and the associated volume fraction of reinforcement when computed to a metal flow rate of 40 g/s and fluidized bed volume of 4.73×10^{-4} and 1.74×10^{-3} m^3 is shown in Fig. 5. From this Figure, it is evident that the volume fraction of reinforcement is generally less than 10% [55].

The geometry of the coaxial injector is shown in Fig. 6 [63]. This type of injector has two coaxial tubes: (i) an injector tube, and (ii) a nozzle tube. A separation distance distances the two tubes. During injection, pressurized gas passes through the gas injector tube into the nozzle tube. The injector tube feeds particles to the nozzle tube by generating an aspiration pressure at the point of separation. The reinforcing ceramic particulates are entrained in the injection gas and accelerated in the nozzle tube. Due to inherent complexity associated with particulate entrainment in the coaxial injectors, reliable theoretical models for mass flow rate are yet to be developed. Experimentally, the mass flow rate can be determined by using the variation in mass of particles in the injector as a function of time:

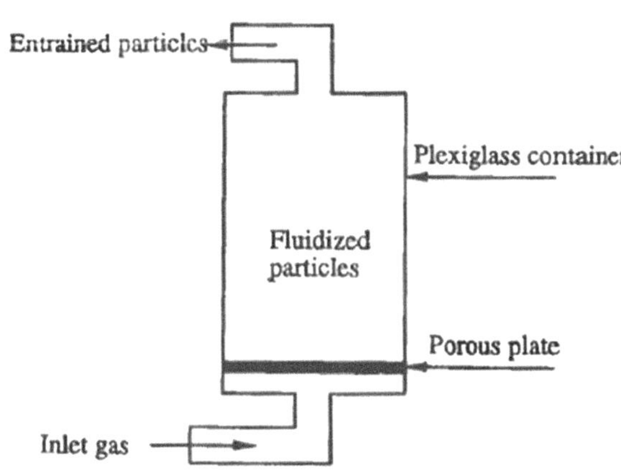

Fig. 4 Schematic diagram showing a fluidized bed injector [63]

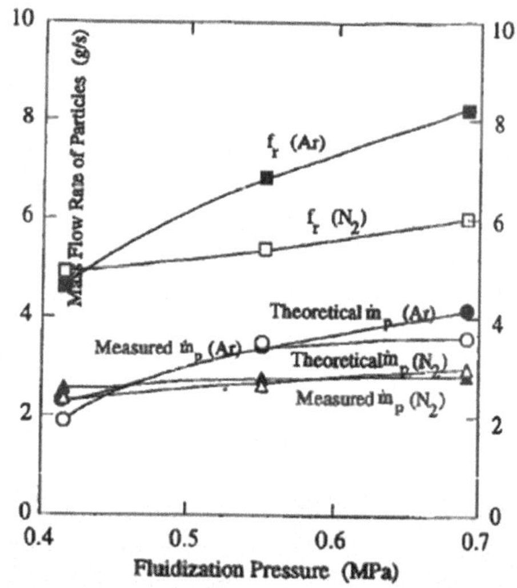

Fig. 5 Measured and computed mass flow rate and volume fraction of SiC particulates as a function of injection gas and fluidization pressure for a fluidized bed volume of 1.74 × 10^{-3} [55]

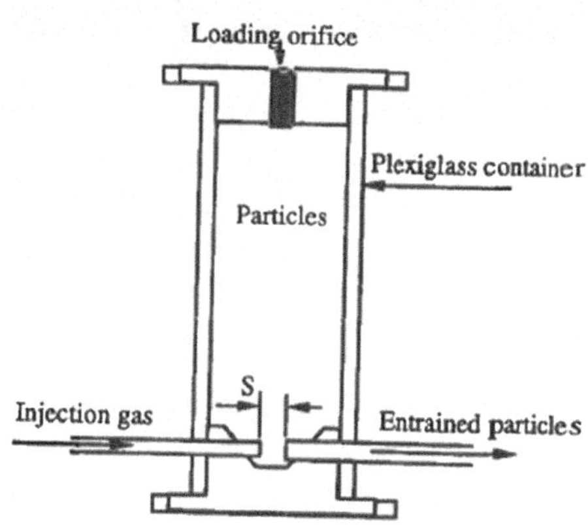

Fig. 6 Schematic showing a coaxial injector with S being the separation distance between the injector tube and the nozzle tube [63]

$$\mathbf{m_p} = \Delta M_p / \Delta t \quad (1)$$

The mass flow rate of the particles increases with increasing injection pressure in the 0.14–0.21 MPa range as shown in Fig. 7. Thereafter, a quasi-steady state is reached and the mass flow rate varies marginally with increasing injection pressure. Also, the mass flow rate of the particles increases with increasing separation distance between the injector tube and the nozzle tube (Fig. 8). It is evident that a quasi-steady flow rate is attained at a separation distance greater than 0.76 cm. Also, the mass flow rates obtained using the coaxial type of injector are in the range of 30–60 g/s for injection pressures in the range of 0.1–0.5 MPa, which is greater than those associated with fluidized bed injectors [less than 10 g/s for injection pressures in the range of 0.4–0.7 MPa].

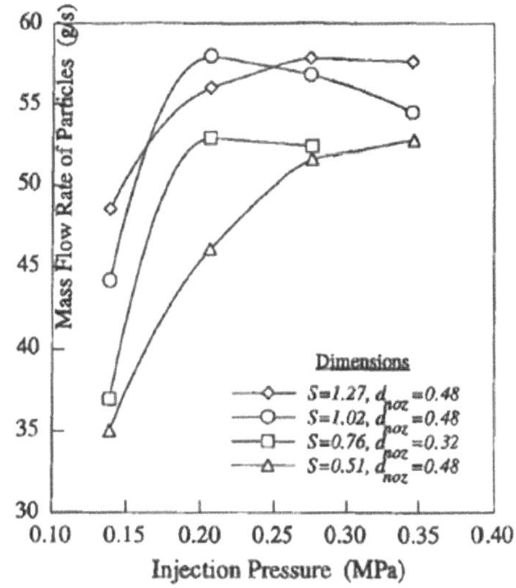

Fig. 7 Mass flow rate of particles as a function of injection pressure measured for a coaxial, injector where S is the separation distance and d is diameter of nozzle of the injector tube (the diameter of the nozzle used is 1.27 cm) [81]

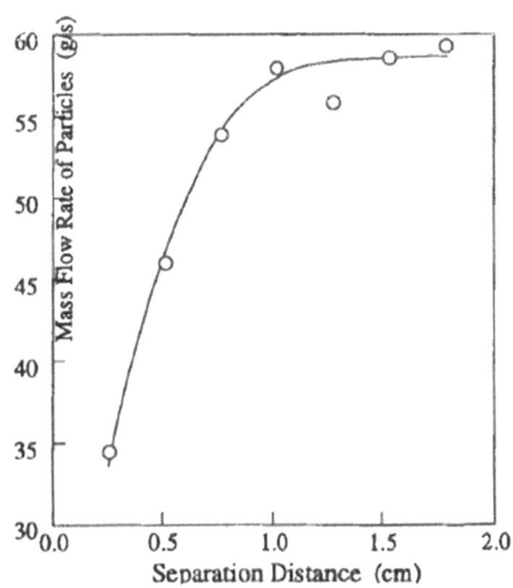

Fig. 8 Mass flow rate of particles as a function of separation distance measured for a coaxial injector for the conditions of injector tube diameter of 0.48 cm, nozzle tube diameter of 1.27 cm and an injection pressure of 0.21 MPa [81]

In the spray atomization and co-injection experiments the reinforcing ceramic particulates were incorporated into the metal matrix while the latter is in semi-liquid state. Few of the advantages specific to this method are the following [46, 52, 63, 82, 83]:

(1) Reduced contact temperature between the ceramic reinforcements and the metal matrix.
(2) Minimized interfacial interactions.

(3) Rapidly solidified microstructures.
(4) Limited segregation of the reinforcement.

Besides, when conditions for incorporation of the reinforcing phase (s) are properly controlled, spray atomization and co-deposition can be used for the synthesis of hybrid or layered MMCs, as well as functionally graded materials (FGMs) [49, 85]. In spray atomization and co-deposition, the ceramic particles are incorporated into the metal matrix by means of using either one injector or more than one gas injector. Hence, in principle it is feasible to synthesize metal matrix composites (MMCs) having more than one type of reinforcing phase. Such matrices reinforced with a mixture of reinforcements are referred to as hybrid composites. The fabrication of hybrid MMCs can be easily realized by injecting mixtures of different types of ceramic particles. Zhang et al. [84] incorporated silicon carbide (SiC) and graphite particle mixtures at volume ratios of 5:5 and 6:4 into an atomized spray of aluminum alloy 6061 by using gas injection during spray atomization and deposition processing. The resultant metal matrix composite (MMC) contained both SiC and graphite as reinforcements with a total volume fraction in the range of 15–27%.

Other related studies have revealed that the technique of spray atomization and co-deposition can be successfully utilized for the manufacture of layered MMCs having an attractive combination of strength and fracture toughness [84, 85]. With specific regard to reinforcement volume fraction an inspection of the results published in the available literature reveals that particle volume fractions ranging from 5 to 25% can be easily achieved. Although volume fractions as high as 30–40% have been reported, these are difficult to reproduce in a reliable manner, and segregation of the particles, throughout the spray deposited matrix, is often observed [86]. As early as 1985, Singer and Ozbek [87] reported that up to 36 vol.% of silicon carbide (SiC), alumina (Al_2O_3), chilled iron, graphite, sand particles, and even healthy mixtures of these could be successfully incorporated into pure aluminum and Al-5Si alloy using the technique of spray atomization and co-deposition. These researchers also observed that all the different particles were distributed homogeneously through the metal matrix, regardless of their differences in density and morphology. In a companion study, Singer reported that the as-spray deposited MMCs could be readily hot rolled to produce composite strips for purpose of examination, analysis, and testing.

Spray Deposition Processing of Pre-mixed Metal Matrix Composites [MMCs]

In spray deposition processing of pre-mixed MMCs the starting materials for the atomization and deposition experiments were composites [49, 50, 53, 62]. The reinforcement phase(s) are introduced by using one of the two methods shown in Fig. 9.

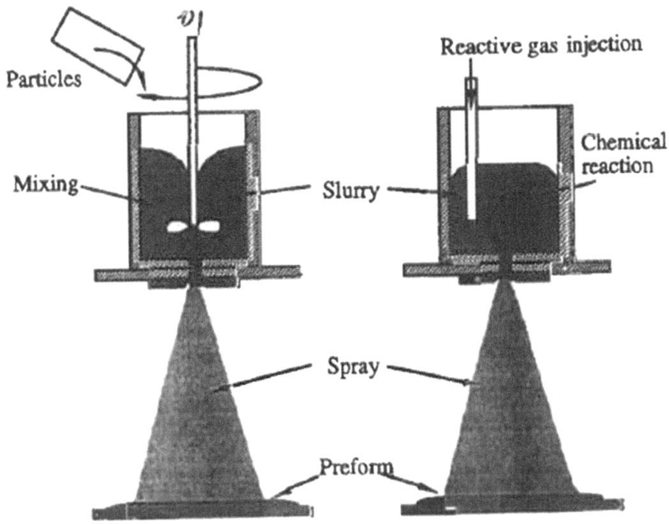

Fig. 9 Schematic diagram showing the spray deposition synthesis of premixed metal matrix composites

Method 1: An in situ formation of the reinforcement phase(s)

In this method, the matrix material is initially melted and the reinforcements are formed in a crucible by chemical reactions between the molten metal and the gaseous or solid phases. The resultant composite slurry is spray atomized and deposited.

Method 2:

A solid metal matrix composite is used as the starting material for purpose of melting and atomization, or, alternatively, the particles and molten matrix are mixed to form a slurry. Subsequently, the slurry is atomized into fine droplets, which are deposited to form a three-dimensional preform. By this method, Gupta et al. [53] successfully synthesized two spray deposited aluminum-lithium based metal matrix composites containing 10 vol% SiC and 5 vol% graphite, respectively.

The primary advantage of spray atomization and deposition of pre-mixed composites is an ability to obtain a near uniform distribution at both the microscopic and macroscopic levels. Using this method, Srivatsan and Lavernia [61] synthesized and characterized a spray deposited AA356/SiCp MMC. In this study, the AA356/SiCp MMC produced using the DURAL method was used as the starting material. During spray deposition, the composite material was re-melted by induction heating and deposited as a preform. In the ingot cast material, an agglomeration of the reinforcing SiC particulate reinforcements was evident resulting in particulate-rich and particulate-depleted regions. The reinforcing silicon carbide (SiC) particles in the cast material were present both at and along the dendrite boundaries forming an inter-dendritic network. However, in the spray deposited material the reinforcing SiCp were distributed uniformly through the aluminum alloy metal matrix. However, there are a few observable disadvantages associated with spray deposition processing of pre-mixed MMCs.

(1) First when the reinforcing phases are introduced by re-melting pre-mixed MMCs they are exposed to liquid matrix material for a prolonged time. Thus, thermal exposure and associated interfacial reactions, depending on relevant reaction kinetics, cannot be avoided. Interfacial reactions do place a limitation on the combination of metal matrix and reinforcement(s) that can be successfully synthesized using this approach.
(2) Fluidity of the composite slurry gradually decreases with an increase in particle volume fraction.

Consequently, for increased reinforcement volume fraction it may be difficult to deliver the composite slurry to the gas atomizer. Therefore, viscosity places a limitation on the volume fraction of reinforcement that can be successfully incorporated into the metal matrix using this method.

Reactive Spray Deposition

In this technique, the particulate reinforcements are introduced into the metal matrix by promoting chemical reactions during both atomization and deposition to produce thermo-mechanically stable dispersoids [69, 70, 76, 78, 88]. Accordingly, this processing technique encompasses two distinct approaches:

(i) Reactive spray atomization and deposition (RSAD), and
(ii) Reactive plasma spray deposition (RPSD).

Based on condition of the reacting phases, reactive spray deposition processes [RPSD] can be classified into one of the three categories:

(A) Gas-liquid interaction,
(B) Liquid-Liquid interactions, and
(C) Liquid-solid interactions.

This is shown in Figs. 10, 11 and 12.

Gas-Liquid Interactions

Gas-liquid interactions represent perhaps the most exhaustively studied area in reactive atomization and deposition. This approach combines atomization, reaction and consolidation into a single step for synthesizing composite materials, as shown in Fig. 10. It offers an opportunity for continuous control of alloy composition and chemical reactions that can occur between the atomized droplets and reactive

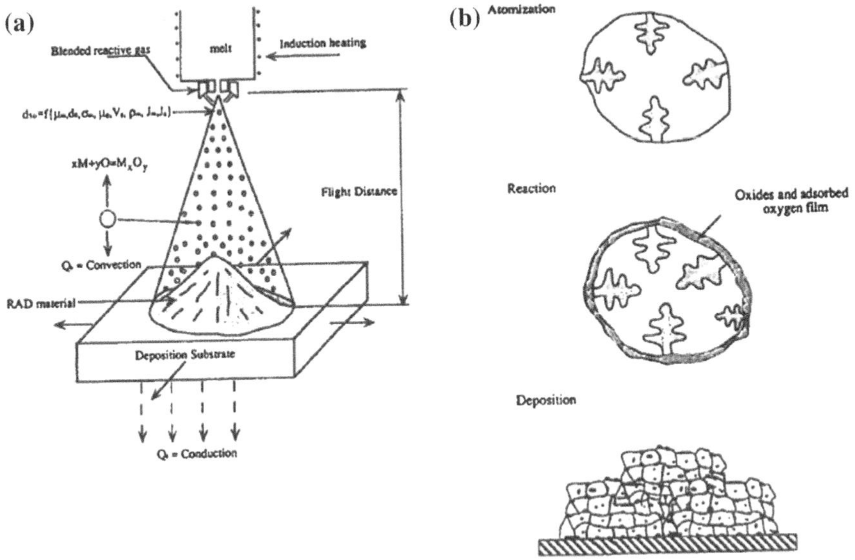

Fig. 10 Schematic diagram showing the reactive spray deposition processes involving liquid-gas reactions: **a** reactive spray deposition apparatus, **b** formation of oxides on the surface [76]

Fig. 11 Schematic diagram illustrating reactive spray atomization and deposition utilizing liquid-liquid reactions

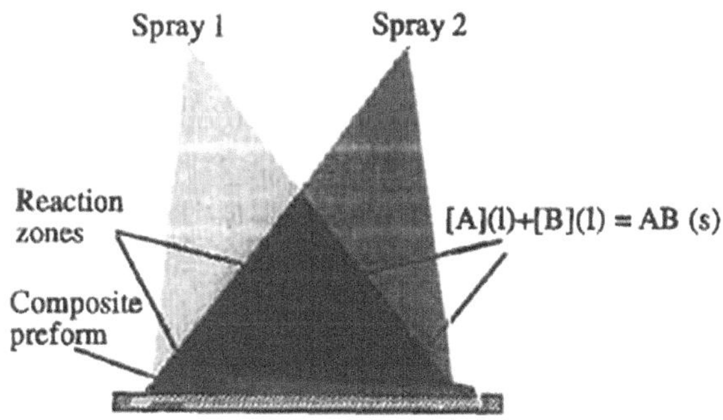

Fig. 12 Schematic diagram illustrating reactive spray atomization and deposition utilizing liquid-solid reactions

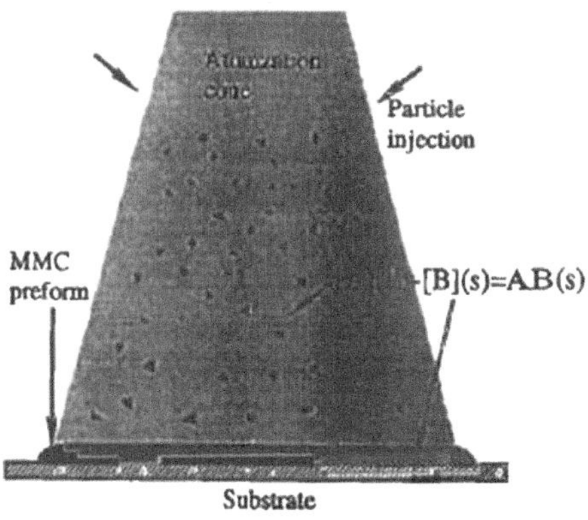

atomization gas. By a judicious selection of alloying additives and reactive gas combinations, based on thermodynamic considerations, it is possible to synthesize materials containing in situ dispersoids, such as: (a) carbides, (b) nitrides, and (c) oxides. During reactive spray atomization and deposition, the molten alloy is atomized using a reactive gas mixture. The droplets are deposited onto a water-cooled copper substrate. Chemical reactions occur between the matrix material and the resistance gas during both atomization and deposition.

To introduce Al_2O_3 dispersoids in a copper matrix Perez and Morris [68] conducted reactive spray deposition experiments on a Cu-1 wt% Al alloy. In this study, nitrogen plus oxygen [$N_2 + O_2$] gas mixture was used for atomization to promote the following oxidation reaction.

$$\text{Cu [Al] (liquid)} + O_2(\text{gaseous}) = \text{Cu (liquid)} + Al_2O_3(\text{solid}) \quad (2)$$

Results revealed that the oxygen and aluminum content in the as-spray deposited preform depends on the level of oxygen concentration in the atomization gas [68]. When the gas mixture contained less than 2 vol% of oxygen gas, its content was not detected in the spray deposited preform. With an increase in oxygen (O_2) concentration in the atomization gas, the oxygen content increased while the aluminum content decreased in the spray deposited preform. For example: a gas mixture containing 8 vol% of oxygen, the volume fraction of alumina or aluminum oxide (Al_2O_3) calculated from the oxygen content was only 1%, and the gas mixture containing 15 vol% of oxygen, the volume fraction of alumina (Al_2O_3) decreased to 0.04% due to a significant loss of aluminum during spray deposition.

Researchers have also investigated the reactive spray deposition behavior of steels to promote the formation of Al_2O_3 particles as dispersoids through the following oxidation reaction [69, 70]

$$\text{Fe [Al] liquid} + O_2(\text{gas}) = \text{Fe (Liquid)} + Al_2O_3(\text{solid}) \quad (3)$$

and

$$\text{Fe [Al] (liquid)} + \text{Fe (Liquid)} + O_2(\text{gaseous}) = \text{Fe (Liquid)} + FeAl_2O_4(\text{solid}) \quad (4)$$

The salient advantages associated with reaction spray atomization and deposition synthesis of metal matrix composites are the following:

(a) A fine reinforcement size, and
(b) Near uniform distribution of the reinforcement particle.

A major challenge of this method is appropriate selection of both the matrix material and the reinforcing phases. Zeng et al. [76, 89] provided an insight into the mechanisms that govern the evolution of dispersoids (Al_2O_3) from gas-liquid reactions. These researchers documented the possible steps that are rate limiting:

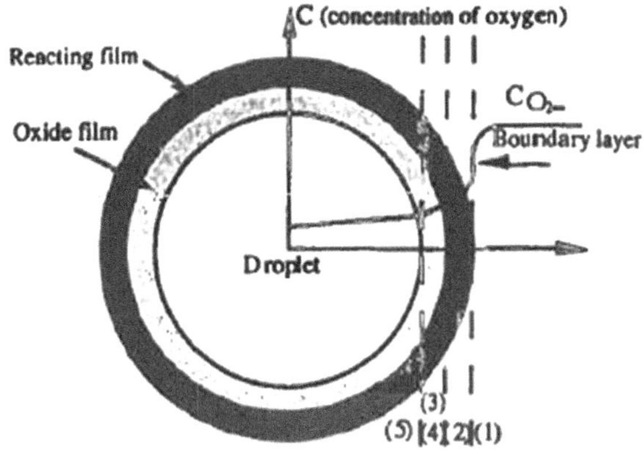

Fig. 13 Schematic showing possible rate limiting steps to reaction between the droplet and oxygen. Were C being the oxygen content in the atomization gas mixture. The numbers in parenthesis refer to the various reaction steps [76]

(1) Mass transfer of oxygen in the gas boundary layer surrounding the droplets.
(2) Adsorption of the oxygen atoms at the surface of the droplets.
(3) Mass transfer of oxygen through the oxide film.
(4) Selective chemical reaction of oxygen with the reactive elements.
(5) Mass transfer of oxygen in the droplets (Fig. 13).

The oxides were found to behave as a diffusion barrier at the surface of the droplet. During the initial stages of atomization, the oxide film is discontinuous. Consequently, the rate of chemical reaction or the rate of diffusion, becomes the rate-limiting step.

Based on the assumption of diffusion within the oxide film being the rate limiting step, Liu et al. [90] formulated a mathematical model to simulate the oxidation kinetics that is present during reactive spray deposition [RSD]. These researchers proposed that when an oxide film is formed on the droplet surface the thermal energy that is released raised the local temperature in the immediate vicinity of the oxide-liquid (or oxide/solid) interface. This necessitates the need for an energy balance condition to be satisfied at the oxide—matrix interface, i.e., the release of heat from oxidation must equal the heat conducted across the interface.

Liquid-Liquid Interactions

Reaction spray deposition using liquid-liquid interactions involve the mixing of two molten metals either prior to or immediately following atomization. This is shown in Fig. 11. This processing technique has still not been explored in detail. In principle, the chemical reactions that occur between two liquid streams produce refractory particle dispersion [78]. Controlling the cooling rate of the droplets and/or the cooling rate of the preform controls the size of the resultant dispersoids. An example of liquid-liquid interactions is provided for copper alloys as [78]:

$$\text{Cu [Ti] (Liquid)} + \text{Cu [B] Liquid)} \rightarrow \text{Cu (liquid)} + \text{TiB}_2\text{(solid)} \tag{5}$$

The alloying element titanium and boron from two different sprays of copper react resulting in the formation of TiB_2 dispersoids. The results obtained using this approach to produce copper alloys having an extremely fine dispersion of TiB_2 particles were encouraging [91].

Liquid-Solid Interactions

Reactive spray atomization and deposition [RSAD] using liquid-solid interactions is shown in Fig. 12. In this processing technique, solid particles of one of the reactants are injected into the atomized spray of a matrix. The reinforcing particles react with the metal matrix both during atomization and deposition to produce reinforcement phases. The process has been used to produce stainless steels containing TiN particle dispersions [78, 88, 92]. Stainless steels containing 2 wt% titanium were spray deposited with Cr_3N_2 or Cr_XN co-injection. During spray deposition, the chromium nitride particles react with the metal matrix to produce TiC particles according to the following reaction:

$$\text{Ti (liquid)} + \text{Cr}_x\text{N (solid)} = \text{Ti [Cr]} + \text{TiC (solid)} \tag{6}$$

Available experimental results reveal that stainless steels containing up to 2 vol% of TiC particles may be easily produced using this method [78, 88, 92].

Liquid-solid reactions have also been used to produce TiC particles in Fe-5 wt% Ti alloy [93]. The results of Majagi et al. [93] revealed that the Fe-5 wt% Ti produced by reactive spray deposition [RSD] with liquid-solid reactions had an equiaxed microstructure with no evidence of co-injected Fe-C particles. The fine precipitates, smaller than 5 μm in size, present both at and along the grain boundaries were identified to be TiC and $FeTi_2$.

Plasma Reactive Spray Deposition

The detailed illustration of a reactive plasma spray source and the deposition process is shown in Fig. 14 [94]. In this method, metallic particles are injected into a plasma source whereby they are heated and re-melted into a spray of liquid and semi-liquid droplets [40]. To produce reinforcing phases a reactive gas is introduced into the plasma source. The metallic droplets react with reactive gas during plasma spraying. Finally, the droplets containing the reaction products are deposited either onto a collecting substrate or a shaped container to form a three-dimensional preform. This method permits the synthesis of a wide variety of metallic, intermetallic and ceramic matrix composites [95–98]. Important examples are carbide particle reinforced MMCs, such as tungsten-tungsten carbide composites, titanium-titanium carbide

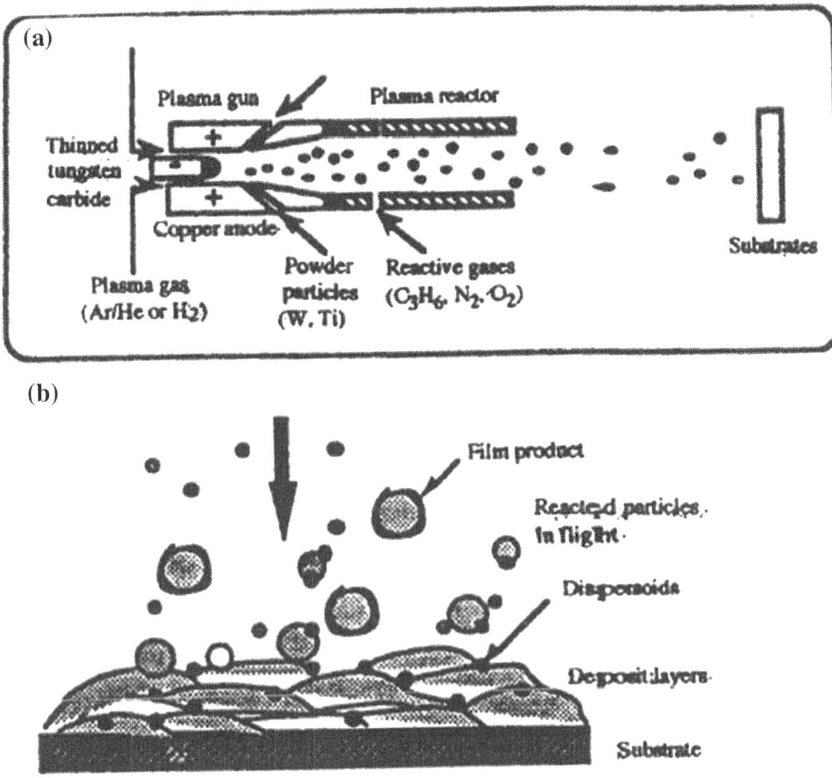

Fig. 14 Schematic diagram illustrating the reactive plasma spray deposition process. **a** Reactive plasma source, **b** Formation of composite preform [94]

composites [95], and $MoSi_2$–SiC composites [96, 97]. In most cases the hydrocarbons react with the metallic elements in the matrix to produce carbides. Tiwari et al. [95] studied the reactive plasma behavior of tungsten-tungsten carbide composites. They used an induction plasma torch to re-melt elemental tungsten powders. To produce the tungsten carbide reinforcements from chemical reactions they used the hydrocarbons as feedstock material. During plasma spraying, the hydrocarbons decomposed and react with the elemental tungsten droplets to form tungsten carbide. The spray of droplets containing tungsten carbide is subsequently deposited onto a deposition surface to form the bulk material.

Incorporation of Reinforcing Phase(s) During Co-injection

There are several approaches that have been successfully utilized to incorporate the reinforcing particulates during spray atomization and deposition processing. An inspection of the relevant literature reveals that the most widely used approach is based on co-injection of a stream of inert gas with entrained particles into the atomized droplet spray (see Fig. 1). The problems posed by an incorporation of a distribution of ceramic particles into the metal matrix by co-injection can be analyzed in the context of a generalized gas-to-metal transition. In liquid phase processing of the MMCs, gas injection or mechanical stirring has been used to incorporate ceramic particles into molten metal matrices [99, 100]. The gas-to-metal transition of ceramic particles is also encountered in surface modification technologies in which the surface of a metallic material is partially melted using a high-energy source, such as: (a) laser beam, or (b) an electron beam [101–103]. Simultaneously, ceramic particles are incorporated into the molten metal pool by means of gas injection to produce a layer of abrasive resistant metal matrix composite.

Incorporation Behavior of Reinforcements

Available experimental evidence suggests that there are several interaction mechanisms that govern the incorporation of ceramic particles into metal matrices using the technique of spray atomization and co-deposition processing. These mechanisms are shown in Fig. 15, and include the following [54]:

(a) Penetration of the ceramic particles into the droplets at the deposition surface,
(b) Mechanical entrapment of the particles during impingement of the droplets, and
(c) Penetration of ceramic particles into the droplets during co-injection.

Since the injected particles lack sufficient kinetic energy to penetrate the droplets they tend to adhere to the surface of the droplets. However, the surface particles may eventually become completely embedded in the metal matrix if there exists sufficient energy during impingement. Also, the particles may reside at the prior

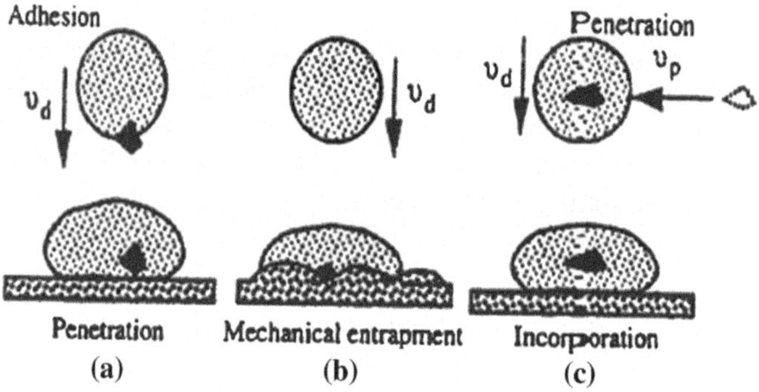

Fig. 15 Incorporation of ceramic particles during spray atomization and co-deposition. **a** Penetration during deposition, **b** Mechanical entrapment during deposition, **c** Penetration during co-injection, where V_d is droplet velocity and V_p is velocity of particle [54]

droplet boundaries and be mechanically entrapped by the droplets that impinge on the deposition surface at a later time. Besides, the ceramic particles may penetrate the atomized droplets when they possess sufficient kinetic energy relative to the droplet [54].

The interactions that tend to occur between the co-injected ceramic particles and atomized droplets are dependent on the conjoint and mutually interactive influences of the following factors:

(i) Atomization and co-injection parameters,
(ii) Droplet solidification condition,
(iii) Physical properties of the particles and those of the droplets, and
(iv) Wetting behavior between the reinforcing particles and droplets of the matrix.

Regarding penetration behavior Wu and Lavernia [59] conducted a systematic study on the Al–Si/SiC system. To that effect, they atomized an Al-4 wt% alloy into droplets and co-injected the SiC particles during atomization. In this study, the deposition substrate was removed to facilitate collection of the composite droplets produced during atomization and ceramic co-injection. To provide an insight into the factors that affect the penetration behavior of the ceramic particles, Wu and Lavernia [59] also examined the distribution and volume fraction of SiC particles in the droplets as a function of size of the droplet. They observed the distribution of ceramic particles in the atomized droplets to depend on size of the droplet. The droplets that were smaller than 98 μm, the reinforcing SiC particles are relatively uniformly distributed. However, for the larger droplets (greater than 180 μm in size) the SiC particles were observed to be segregated to the droplet surface [59]. For droplets approaching a size of 30 μm the reinforcing SiC particles were found to be present primarily at the droplet surfaces. Wu and Lavernia [59] also determined the volume fraction of the SiC particles present in the atomized droplets. Their results revealed the volume fraction as a function of droplet size to exhibit a maximum value for the intermediate size droplets [example: 98 μm for the Al-4 wt % Si/SiC system]. Accordingly, both an increasing droplet size and decreasing droplet size lead to a concurrent decrease in the number of ceramic particles incorporated in the droplets during co-injection.

The penetration behavior of the co-injected particles, is shown in Figs. 16 and 17, is rationalized by considering solidification behavior of the droplet. This necessitates the need for a comparison to be made between the droplet temperature and the liquidus temperature at the injection distance and assuming no solidification. Accordingly, if temperature of a droplet is greater than the liquidus temperature the droplet may be assumed to be completely molten. If the temperature falls below the liquidus temperature the droplet may be in one of the three states: (i) under-cooled liquid, (ii) partially solid, and (iii) completely solid.

From an analysis of the solidification characteristics of the droplets at the injection distance, Wu and Lavernia [59] proposed two mechanisms for the observed distribution and volume fraction of SiC particles in the droplets. These are the following:

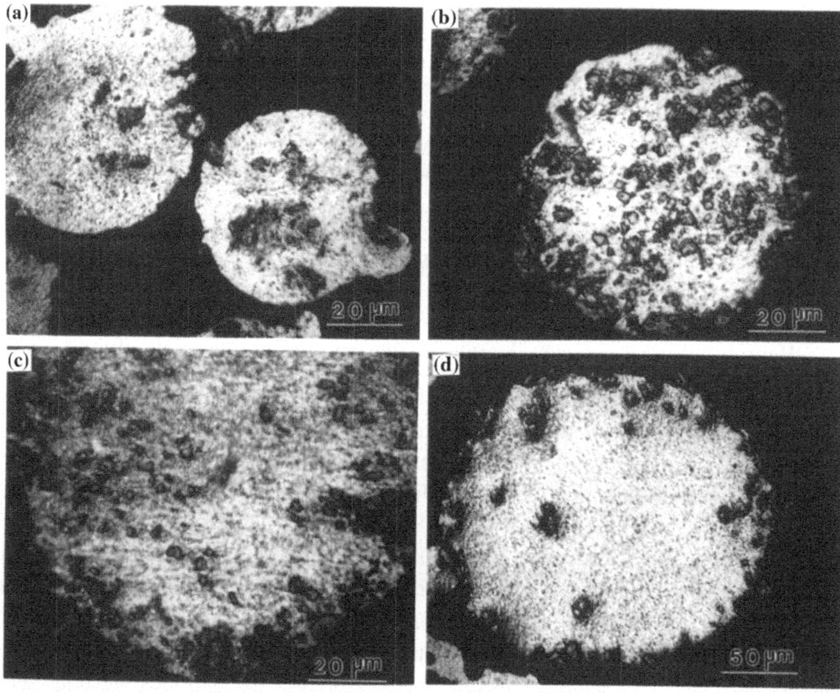

Fig. 16 Optical micrographs showing the distribution of silicon carbide particles in Al-4 wt% Si droplets of various diameter: **a** d = 43 μm, **b** d = 83 μm, **c** d = 116 μm, and **d** d = 230 μm [59]

Fig. 17 Volume fraction of SiC particles in Al-4 wt.% Si droplets as a function of size of the droplet [59]

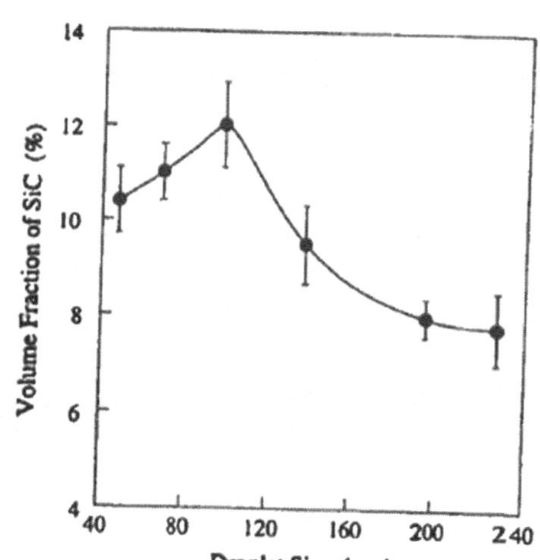

(a) Penetration of the reinforcing SiC particles into the droplets, and
(b) Segregation of the reinforcing SiC particles during penetration.

Segregation of the reinforcing SiC particles to the surface of the droplets during solidification implied that to attain a homogeneous distribution of particles, penetration alone is not sufficient Accordingly, an interaction of the particles with the solidification front, at the droplet interior, must be taken into consideration.

Regarding the effect of particle type on incorporation behavior during spray atomization and co-injection, available experimental results reveal that penetration is critically dependent on both the physical and chemical characteristics of the particles and the droplets. For example, a relatively large number of ceramic particles were found to be present throughout the cross-section of Al–Si/SiC and Al–Si/TiB_2 droplets. However, in the aluminum/graphite system, the reinforcing graphite particles were found to be present primarily at the droplet surface. Wu et al. [81] attributed these experimental observations to intrinsic differences in density and wetting behavior among the various types of particles. Accordingly, a large particle density coupled with a low wetting angle with the liquid would facilitate ease in penetration [81]. The differences in penetration behavior associated with various reinforcement particles have also been noted by other investigators working on the spray atomization and co-deposition synthesis of composites [66].

Penetration of the Semi-liquid Droplets

To enhance our understanding of the fundamental mechanisms that govern droplet penetration during co-injection, researchers working in this area have proposed several physical models [59, 67, 81]. The penetration of particles can take place either during collision with the atomized droplets or upon impingement with the deposition surface. In both cases, the particles are at a relative velocity with the droplets. Injection of ceramic particles into a molten alloy is shown in Fig. 18 [104]. In this process, the ceramic particles are accelerated using a carrier gas and

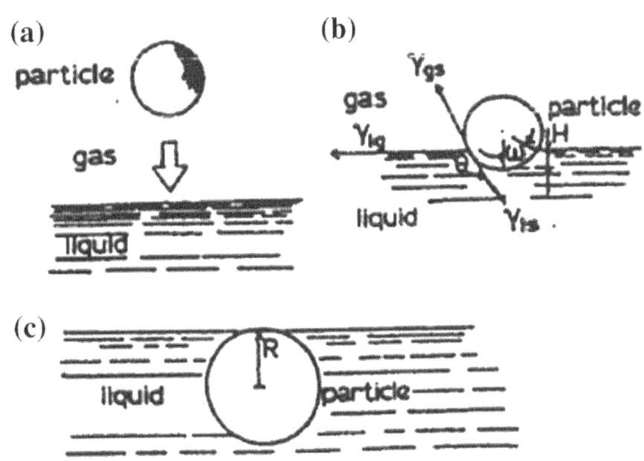

Fig. 18 Steps involved in the transfer of a particle from a gas to a liquid. **a** Before penetration, **b** Particle liquid surface, and **c** After penetration. R is radius of the particle. Theta is wetting angle and H is the penetration distance, and γ_{gs}, γ_{lg} and γ_{ls} are the gas-solid, liquid-gas, and liquid-solid surface energy [100]

are directed onto the surface of a molten alloy. When the injection velocity is high, the ceramic particles tend to penetrate the liquid surface resulting in incorporation. The particle penetration behavior encountered during spray deposition and gas injection onto a liquid surface was observed to be analogous in that both processes involve dynamic transfer of the reinforcement particles from a gaseous phase to the matrix material. The notable difference being that in spray deposition processing the matrix material is in a semi-liquid state.

Penetration of the Solid Droplets

For a particle to penetrate a solid droplet, impingement of the particle must result in plastic deformation of the droplet surface. The problem concerning penetration behavior of a hard-ceramic particle into a relatively soft metallic surface has been addressed by Tiwari and Herman [105]. The dynamic pressure (P) exerted by a ceramic particle on a solid droplet can be calculated using a simple mathematical expression, since the elastic modulus of a ceramic material is much higher than that of a metallic counterpart [105].

$$P = 0.558\left[\rho_p\{E/\left(1-\upsilon_p^2\right)^4\}\right]^{0.2} V^{0.4} \tag{7}$$

where E and υ_p are elastic modulus and Poisson's ratio of the droplet, ρ_p is particle density, and V is relative velocity between the reinforcing particle and the droplet. The depth of crater (d_c) caused by the impingement can be expressed as [105]:

$$d_c = \left[0.14\,m_p V^2/\pi\left(d_p/2 - d_c/3\right)\sigma_{ym}\right]^{0.5} \tag{8}$$

where σ_{ym} is yield strength of the droplet material. Wu and Lavernia [59] proposed that penetration of the reinforcing SiC particle is possible only when depth of the crater is greater than size of the reinforcing particle. If not, the reinforcing particle would tend to adhere to the surface.

Evolution of Microstructure

The spray deposited metal matrix composites exhibit microstructural features that are similar to those present in as-spray deposited monolithic materials, such as: (a) equiaxed grain morphology [41, 47, 48, 51, 66, 73, 106], (b) phase modification [52], (c) extended solid solubility [63], and (d) limited porosity [40, 41, 52]. The differences in microstructural features between the spray deposited metal matrix composites and the monolithic materials, as induced by the presence of one or more reinforcing phases is briefly highlighted.

Distribution of the Reinforcing Phase

The microstructure of spray deposited metal matrix composites generally exhibits significantly less segregation of the reinforcements relative to materials processed utilizing conventional ingot solidification techniques. In Fig. 19 is shown the microstructure of 6061 aluminum alloy/graphite MMC and 6061 Al/SiC MMC synthesized by the technique of spray atomization and co-deposition [47, 107]. When compared one-on-one with the microstructure of the ingot cast MMCs the spray deposited MMCs exhibit fewer particle-particle contacts and a relatively uniform distribution of the reinforcing particles. These observations agree with those put forth by other investigators [41, 42, 52, 54, 108]. However, high magnification observation of the microstructure revealed some degree of segregation at

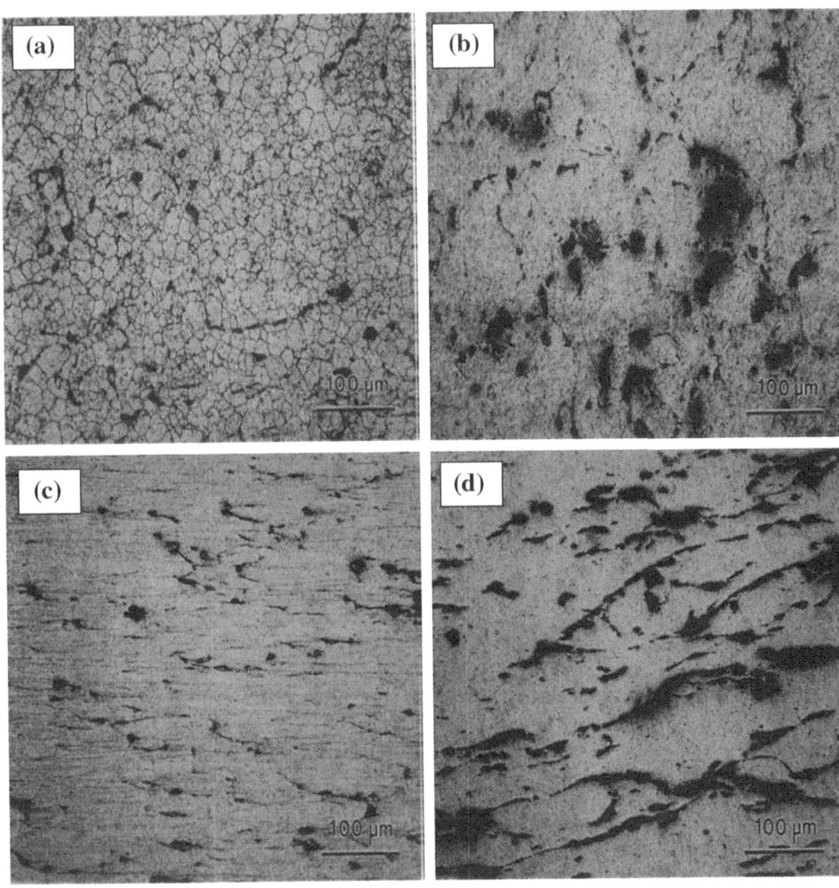

Fig. 19 -Optical micrographs of spray deposition processed MMCs: **a** Grain morphology of the as-spray deposited 6061 aluminum alloy with 0.05 graphite MMC, **b** As-spray deposited 6061 aluminum alloy 6061 with 0.10 graphite MMC, **c** Extruded aluminum alloy 6061/0.05 graphite MMC, **d** Extruded aluminum alloy 6061/0.10 graphite MMC. From: [107]

Fig. 20 Optical micrograph showing microstructure of spray atomized and deposited 6061 Al-SiC_P

the fine microscopic level. An example of such microscopic segregation is shown in Fig. 20 for a spray deposited 6061 Al/SiC metal matrix composite reinforced with 10 vol% of SiC particles. The micrograph reveals the presence of reinforcement-rich and reinforcement depleted regions [82, 83]. This observation accords well with the results of White and co-workers [42] who also observed limited reinforcement segregation in an 8090/SiC and 8090/B4C metal matrix composites synthesized by the technique of spray atomization and co-deposition. The non-uniform distribution of the reinforcing phase can be attributed to two possible causes [59]:

(a) An absence of entrapment by the solid/liquid interface, the particles will be pushed towards the last zones to solidify, which are generally the grain boundaries. Thus, the resultant inter-particle spacing depends on grain size, which is linked to the thermal environment during deposition.
(b) The presence of pre-solidified powders during co-injection.

Under typical injection conditions used in spray atomization and co-deposition the particles will not penetrate the solid powders. Should the powders retain their spherical morphology upon impingement with the deposition surface (i.e., they do not experience re-melting), the size and/or dimensions of the particle depleted regions will be limited by size of the pre-solidified powders.

Presence of an Interface

The available ceramic reinforcements are highly reactive with the molten metal at an elevated temperature. Consequently, prolonged contact between the reinforcing phase and the molten metal, encountered during liquid phase processing of metal-matrix composites, generally leads to an extensive interfacial reaction. For example, in the Al/SiC system brittle intermetallic phases, such as: Al_4C_3 and Al_4SiC_4, tend to form at the reinforcement-matrix interfaces either as a continuous layer or as isolated precipitates. Reactions between liquid aluminum and the reinforcing SiC particles may be exacerbated in accordance with the following reaction [109]:

$$
\begin{aligned}
&4\,\mathrm{Al}\ (\text{liquid}) + 3\ \mathrm{SiC}\ (\text{solid}) = \mathrm{Al_4C_3}(\text{solid}) + 3\,\mathrm{Si} \\
&\text{and} \\
&4\,\mathrm{Al}\ (\text{liquid}) = 4\ \mathrm{SiC}\ (\text{solid}) = \mathrm{Al_4SiC_4}(\text{solid}) + 3\,\mathrm{Si}
\end{aligned}
\tag{9}
$$

The first reaction proceeds by solid state diffusion of carbon through Al_4C resulting in a continuous reaction layer. The second reaction occurs by the dissolution of SiC into molten aluminum. The reaction between aluminum matrix and graphite occurs in conformance with the following relationship [110–113].

$$4\,\mathrm{Al}\ (\text{liquid}) + 3\,\mathrm{C}\ (\text{solid}) = \mathrm{Al_4C_3}\ (\text{solid}) \tag{10}$$

Although limited interfacial reactions may enhance the load carrying capability of the interface, extensive interfacial reactions are detrimental to mechanical properties.

An absence of interfacial reactions in spray deposited MMCs has been attributed to relatively low contact temperatures coupled with the short contact time between the liquid phase and the reinforcements [52, 63]. In spray atomization and co-deposition, the range of temperatures over which the metallic matrices and ceramic particles are placed in contact with each other is generally between the liquidus temperature and the solidus temperature of the matrix material. This compares well with the contact temperature, which is generally above the liquidus in the case of liquid phase processing. The contact time between the reinforcement and the semi-liquid matrix can be readily estimated from the local solidification time during spray deposition. Even under extreme conditions, i.e., when the spray of droplets contains 49% of liquid the solidification time is less than 220 s [112]. The short contact time may be deduced from a careful observation of the spray-deposited microstructure.

To provide an insight into the behavior of the interface during spray atomization and co-deposition, Gupta et al. [54] re-melted a spray deposited Al-2.1 weight pct. Li/SiC MMC by heating this material to a temperature of 870 °C for 60 min in an atmosphere of argon gas. The results revealed that in the as-spray deposited condition the Al/SiC interfaces were free of interfacial reactions. However, extensive interfacial reactions between the aluminum alloy matrix and the reinforcing SiC particles occurred during re-melting. Interfacial reactions between the reinforcing phase and the metal matrix may also occur during prolonged exposure in the solid state. The overall stability of an interfacial reaction is strictly dependent on the driving force for the chemical reaction kinetics involved. Consequently, it is not always possible to avoid formation of interfacial reactions in all systems by using the technique of spray atomization and co-deposition.

The Presence and Role of Porosity

Spray deposited materials generally exhibit a limited amount of porosity quite like that present in the spray deposited monolithic alloys [48, 52, 63, 73, 107, 108]. Also, in spray deposited MMCs the presence of many reinforcement particles often precludes the detection of a chill zone near the substrate surface. Consequently, most characterization studies of porosity in spray deposited MMCs are conducted in the central portion of the preform. Available experimental evidence reveals at least three types of pores to be present in spray deposited MMCs. These are the following [63, 73]:

(i) Gas-filled pores,
(ii) Unfilled interstices, and
(iii) Solidification shrinkage.

Baskin et al. [73] observed the presence of superficial pores in as-spray deposited CoSi reinforced with silicon carbide particles. In their independent study, the pore size was determined to be on the order of 10 μm. Based on size and morphological considerations the authors attributed such pores to gas entrapment during spray deposition. In a related research study, Gupta et al. [63] noted the size distribution of the pores in spray deposited Al–Ti/SiC MMCs appeared to be bi-modal. Accordingly, small pores were of the size of one to 2 μm and were uniformly distributed, whereas the larger pores were in the range of size ten micron-meters [μm]. The large pores exhibited an irregular morphology and were found to be present at the prior droplet boundaries [63]. Consideration of the location and morphology of the large pores it is evident that they are formed due to an insufficient filling of the interstices by the liquid phase during spray deposition. Such interstices may be created ether between the powder particles or between the powder particles and the deposition surface.

The overall amount of porosity present in spray deposited materials was determined using density measurements. In a quantitative study of porosity in spray deposited MMCs, Perez and co-workers [107] defined the porosity percentage (p) to be:

$$P = (1 - f_r - f_m) \qquad (11)$$

where in the above expression f_r is the volume fraction of reinforcements and f_m is volume fraction of the matrix.

Size of the Grains

The as-spray deposited metal matrix composites exhibit an equiaxed grain morphology like that present in spray deposited monolithic alloys [41, 42, 48–51, 55–58, 66]. The injection of ceramic particles results in a reduction in grain size of the engineered or synthesized metal matrix composite relative to those of the unreinforced counterpart. Grain refinement is associated with an incorporation of the ceramic particles in spray deposited materials. The extent of grain refinement reported for spray deposited MMCs, relative to that of the unreinforced counterpart, ranges from 7–66%. For different materials, an incorporation of the same volume fraction of the reinforcement does not lead to the same degree of grain refinement. Therefore, the sensitivity of grain size to processing parameters must be taken into consideration. In a research study, Gupta and coworkers [56] analyzed this phenomenon and proposed grain size refinement that can be achieved through reinforcement co-injection to depend on the competing and mutually interactive influences of reinforcement volume fraction and atomization parameters, such as gas flow rate and metal flow rate. Accordingly, the effect of reinforcement volume fraction can be compared only when the atomization conditions are maintained relatively constant. Such a comparison was made by Gupta and co-workers [56] using an Al-5 wt% Cu alloy reinforced with silicon carbide particles. A similar trend of decrease in grain size with increasing reinforcement volume fraction was also observed by Wu and Lavernia [47] for aluminum alloy 6061 that was reinforced with 3 μm size silicon carbide particles. It was proposed that the observed grain refinement in spray deposited metal matrix composites, relative to the monolithic counterpart, is a result of either one or a combination of three distinct mechanisms brought about by the co-injection of the ceramic particles:

(a) Thermal effects [52, 55, 56].
(b) Solidification effects [52, 56].
(c) Solid state cooling effects [83].

Key Highlights or Conclusions

1. The development and emergence of spray atomization and deposition processing for the manufacture of discontinuously-reinforced metal matrix composites (MMCs) and even intermetallic matrix composites (IMCs) was motivated by the difficulties associated with conventional casting and use of the powder metallurgy approach.
2. The approach offers the advantage of being able to avoid the extreme thermal excursions and concomitant degradation in interfacial and overall mechanical properties.
3. The processing technique also prevents the occurrence of extensive macroscopic segregation, normally associated with conventionally casting processes.

4. Additionally, the technique also eliminates the need to handle fine reactive particulates, as is necessary with powder metallurgy processing.
5. Several aspects related to the fluid, thermal and solidification phenomenon are now well studied and documented and several materials that have been produced by this technique have shown promising mechanical properties in comparison with those produced by the conventional techniques.

References

1. Koss DA, Copley SM (1971) Thermally-induced residual stresses in eutectic composites. Metall Trans 2A, 1557–1560
2. Brown LM, Stobbs WM (1971) The work hardening of copper-silica. The Philos Mag: J Theoret, Exp Appl Phys, 23(185)
3. Divecha AP, Crowe CR, Fishman SG (1977) Failure modes in composites IV. Metallurgical Society of AIME, Warrendale, PA, pp 406–411
4. Divecha AP, Fishman SG, Karmarkar SD (1981) Silicon carbide reinforced aluminum: a formidable composite. J Metals 33(9):12
5. Nair SV, Tien JK, Bates RC (1985) SiC reinforced aluminium metal matrix composites. Int Metals Rev 30(6):285–296
6. Metal matrix composites overview: paper number MMCIAC no. 253, Santa Barbara, California, USA
7. Schoutens JE (1982) Introduction to metal matrix composites, MMCIAC tutorial series no. 272
8. Taya M, Arsenault RJ (eds) (1989) Metal matrix composites: thermomechanical behavior. Pergamon Press, New York, USA
9. Harrigan WC Jr (1991) Metal matrix composites. In: Everett RK, Arsenault RJ (eds) Metal matrix composites: mechanisms and properties. Academic Press, USA, pp 1–15
10. Lin Yaojun, Zhou Yizhang, Lavernia Enrique J (2004) On the influence of in-situ reactions on grain size during reactive atomization and deposition. Metall Mat Trans 35A:3251–3260
11. McDaniel DL (1985) Analysis of stress-strain, fracture and ductility behavior of aluminum matrix composites containing discontinuous silicon carbide reinforcement. Metall Trans 16A:1105–1115
12. Dermarkar S (1986) Met Mater 2:144–147
13. Chawla KK (1987) Composite materials. Springer, New York
14. Srivatsan TS, Sudarshan TS (1993) Rapid solidification technology: an engineering guide. Techonomic Publishing Inc., PA, USA, pp 603–700
15. Hunt WH Jr, Cook CR, Sawtell RR (1991) Cost effective high performance powder metallurgy aluminum matrix composites for aerospace applications, SAE technical paper series 91-0834, Warrendale, PA, USA
16. Hunt WH Jr (1991) Cost effective high performance composites. In: Presented at international conference on powder metallurgy aerospace materials, Lausanne, Switzerland
17. Clyne TW, Withers PG (1993) An introduction to metal matrix composites, Cambridge Solid State Science Series, Cambridge University Press, UK
18. Kreider KG (1974) Metal matrix composites, composite materials, vol 4. Academy Press, New York, p 1
19. Kelley P (1979) Composites, vol 10, p. 2.t
20. Srivatsan TS, Sudarshan TS, Lavernia EJ (1995) Processing of discontinuously-reinforced metal matrix composites by Rapid Solidification. Prog Mat Sci 39(4–5):317–409
21. Fishman SG (1986) J Metals 38(3):26
22. Flom Y, Arsenault RJ (1986) Mat Sci Eng 77:191

23. Chou TW, Kelly A, Okura A (1985) Composites 16:187
24. Piggot MT (1980) Load bearing fiber composites. Pergamon Press, New York
25. Broutman LJ, Krock RH (1969) Modern composite materials. Addison Wesley Publishing Company, New York
26. Arsenault RJ (1984) Mat Sci Eng 64:171
27. Crow CR, Gray RA, Hasson DF (1985) Proceedings of the fifth international conference on composite materials. In: Harrigan WC Jr, Strife J, Dhingra AK (eds) The metallurgical society of AIME, Warrendale, p 843
28. Ibrahim IA, Mohamad FA, Lavernia EJ (1991) J Mat Sci 26:1137
29. Crowe CR, Hasson DF (1992) Strength of metals and alloys. In: Gifkins RC (ed) Proceedings of the sixth international conference, Pergamon Press, Australia
30. Rack HJ (1988) Adv Mat Manuf Process 3(3):327
31. Srivatsan TS, Ibrahim IA, Mohamad FA, Lavernia EJ (1991) J Mat Sci 27
32. Davidson DL (1987) Metall Transac 18A:2125
33. Logsdon WA, Liaw PK (1986) Eng Fract Mech 24:737
34. Manoharan S, Lewandowski JJ (1990) Acta Metall 328(3):489
35. Davidson DL (1989) Eng Fract Mech 33(6):965
36. Davidson DL (1991) The effect of particulate SiC on fatigue crack growth in a cast-extruded aluminum alloy composite. Metall Trans 22A:97
37. Shang JK, Ritchie RO (1989) Crack bridging by uncracked ligaments during fatigue crack growth in SiC-reinforced aluminum-alloy composites. Metall Trans 20A:897
38. Singer ARE (1972) J Inst Metals 100:185
39. Willis J (1988) Metals Mat 4:485
40. Buhrmaster CL, Clark DE, Smart HO (1988) J Metals 40:44
41. Kojima KO, Lewis RE, Kaufman MJ (1989). In: Sanders TH Jr, Starke EA Jr (eds) Aluminum-lithium alloys V. MCE Publishers, Birmingham, United Kingdom, p 85
42. White J, Palmer IG, Hughes IR, Court SA (1989) Aluminum-Lithium Alloys V. In: Sanders TH Jr, Starke EA Jr (eds). MCE Publishers, Birmingham, United Kingdom, p 1635
43. Llorca J, Ruiz J, Healy JC, Elices M, Beevers CJ (1994) Mat Sci Eng A 185:1
44. Gupta M, Mohamad FA, Lavernia EJ (1989). In: Mostaphaci H (ed) Proceedings of the 17th international symposium on advances in processing and characterization off ceramic metal matrix composites. Pergamon Press, Oxford, UK, p 236
45. Gupta M, Mohamad FA, Lavernia EJ (1990) Solidification behavior of Al-Li-SiCp metal matrix composites processed using variable co-deposition of multi-phase materials. Mat Manuf Process 5(2):165–196
46. Ibrahim IA, Mohamad FA, Lavernia EJ (1991) Particulate-reinforced metal matrix c composites, a review. J Mat Sci 26:1137–1156
47. Wu Y, Lavernia EJ (1991) Spray atomized and co-deposited 6061 Al/SiCp composites. J Metals 43(8):16–23
48. Zhang J, Perez RJ, Gupta M, Lavernia EJ (1993) Damping behavior of particulate-reinforced 2519 aluminum metal matrix composites. Scr Metall Mater 28:91–96
49. Wu Y (1994) Master of science thesis, University of California at Irvine, Irvine, CA
50. Wu M, Srivatsan TS, Pickens JR, Lavernia EJ (1992) Microstructure and mechanical properties of spray deposited Al-Cu-Li-Ag-Mg-Zr alloy. Scr Metall Mater 27:761–766
51. Gupta M, Srivatsan TS, Mohamad FA, Lavernia EJ (1993) Microstructural evolution and mechanical properties of SiC/Al_2O_3 particulate reinforced spray deposited metal matrix composites. J Mat Sci 28:2245–2259
52. Kim NJ, Park WJ, Ahn S, Elias L (1994). In: Upadhyaya K (ed) High performance metal matrix and ceramic matrix composites. The Minerals, Metals and Materials Society, Warrendale, PA, p 137
53. Gupta M, Bowo K, Lavernia EJ, Earthman JC (1993) Effect of particulate type on fatigue crack propagation in Al-Li based spray deposited composites. Scr Metall Mater 28:1053–1058
54. Gupta M, Ibrahim IA, Mohamad FA, Lavernia EJ (1991) J Mat Sci 26:6673

55. Gupta M, Mohamad FA, Lavernia EJ (1991) The effects of discontinuous reinforcements on the distribution of ceramic reinforcements during spray atomization and co-deposition. Int J Rapid Solidification 6:247–284
56. Gupta M, Mohamad FA, Lavernia EJ (1992) The effect of ceramic reinforcements during spray atomization and co-deposition of metal matrix composites: part i; heat transfer. Metall Trans A 23A:831–843
57. Perez RJ, Zhang J, Lavernia EJ (1992) Strain amplitude dependence of 6061 Al MMCs. Scr Metall Mater 27:1111
58. Zeng X, Lavernia EJ (1992) Interfacial behavior during spray atomization and co-deposition. Int J Rapid Solidification 7:219–243
59. Wu Y, Lavernia EJ (1992) Interaction mechanisms between ceramic particles and atomized metallic droplets. Metall Trans A 23A:2923–2937
60. Wu Y, Lavernia EJ (1994). In: Ravi VA, Srivatsan TS, Moore JJ (eds) Processing and fabrication of advanced materials III. The Minerals, Metals and Materials Society Warrendale, PA, p 501
61. Srivatsan TS, Lavernia EJ (1994) Cyclic strain resistance of a spray atomized and deposited cast aluminum alloy metal matrix composites. Compos Eng Int J 4(4):459–472
62. Gupta M, Lane C, Lavernia EJ (1992) Microstructure and properties of spray atomized and deposited Al-7Si/SiCp metal matrix composites. Scr Metall Mater 26:825–830
63. Gupta M, Juarez-Islas J, Frazier WE, Mohamad FA, Lavernia EJ (1992) Microstructure, excess solid solubility and elevated temperature behavior of spray atomized and co-deposited Al-Ti-SiC. Metall Trans B 23B:719–736
64. Armanie KP, Zaidi MA (1987) Metallographic examination of osprey MMC sample, Technical report, Alcoa Technical Center, Pittsburgh, PA, p 1
65. Maker PP, Cantor B, Katerman L (1990). In: Kahn T, Effenberg G (eds) International conference on advanced aluminum and magnesium alloys. ASM International, Materials Park Ohio, p 659
66. Mathur P, Kim MH, Lawley A, Apelian D (1990) Powder metallurgy: key to advanced materials technology. ASM International, Materials Park, Ohio, p 55
67. Majagi SI, Ranganathan K, Lawley A, Apelian D (1992). In: Lavernia EJ, Gungor MN (eds) Microstructural design by solidification processing. The Minerals, Metals and Materials Society, Warrendale, PA, p 139
68. Perez JF, Morris F (1994) Scripta Metallurgica Materialia 31:231
69. Unigame Y, Lawley A, Apelian D (1993) In-situ spray casting of dispersion strengthened alloys I: thermodynamic and reaction kinetics, 1993 powder metallurgy world congress, paper number 14E-T6-6, Kyoto. Japan Powder Metallurgy Association, Japan
70. Ranganathan K, Lawley A, Apelian D (1993) In situ casting of dispersion strengthened alloys-II, experimental studies, paper number 14E-T6-6, 1993 powder metallurgy world congress, Kyoto. Japan Powder Metallurgy Association, Japan
71. Zhao YY, Grant PS, Cantor B (1993) J Microsc 169(2):263
72. Zhao YY, Grant PS, Cantor B (1685) J Physique 3:1685
73. Baskin D, Wolfenstein J, Lavernia EJ (1994) Elevated temperature mechanical behavior of CoSi and particulate reinforced CoSi produced by spray deposition. J Mater Res 9(2):362–371
74. Jeng YL, Wolfenstein J, Lavernia EJ (1993) Scripta Metallurgica Materialia 28:453
75. Jeng YL, Lavernia EJ, Wolfenstein J, Bailey DE, Sickinger A (1993) Creep behavior of plasma sprayed SiC reinforced MoSi2. Scr Metall Mater 29:107–111
76. Zeng X, Liu H, Chu MG, Lavernia EJ (1992) An experimental investigation of reactive spray atomization and deposition processing of Ni_3Al/Y_2O_3 using N_2-O_2 atomization. Metall Trans A 23A:3394–3399
77. Liang X, Lavernia EJ (1992) Interface behavior in Ni_3Al/TiB_2 intermetallic matrix composites. Mat Sci Eng A 153 1–2:654–661
78. Lawley A, Apelian D (1994) Powder Metall 37:123

79. Lawrynowicz DE, Lavernia EJ (1994) Spray atomization and deposition processing of fiber-reinforced intermetallic matrix composites. Scr Metall Mater 31(9):1277–1281
80. Lawrynowicz DE, Lavernia EJ (1995) A review of sensors and techniques used to monitor processing parameters during spray atomization. J Mat Sci 30:1125–1138
81. Wu Y, Zhang J, Lavernia EJ (1994) Modeling of the incorporation of ceramic particulates in metallic droplets during spray atomization and co-injection. Metall Mat Trans B 25 B:135–147
82. Ibrahim IA, Mohamed FA, Lavernia EJ (1990). In: Kahn T, Effenberg G (eds) International conference on advanced aluminum and magnesium alloys. ASM International, Materials Park, OHIO, p 745
83. Lavernia EJ (1991) SAMPE Q 22:2
84. Zhang J, Perez RJ, Lavernia EJ (1994). In: Chawla KK, Liaw PK, Fishman SG (eds) High performance composites. The Minerals, Metals and Materials Society, Warrendale, PA, USA, p 361
85. Wu M, Zhang J, Hunt WH Jr, Lewandowski JJ, Lavernia EJ (1996) J Mat Synth Process, 4:127
86. Singer ARE (1991) Metal matrix composites made by spray forming. Mat Sci Eng 135A:13
87. Singer ARE, Ozbek S (1985) Metal matrix composites produced by spray deposition. Powder Metall 28(2):72
88. Leatham AG, Ogilvy A, Elias L (1993) The osprey process: current status and future possibilities. P/M in Aerospace, Defense and Demanding Applications, San Diego, 7–10 Feb 1993
89. Zeng X, Nutt SR, Lavernia EJ (1995) Microstructural characterization of Ni_3Al processing by reactive atomization and deposition. Metall Mat Trans 26A:817–828
90. Liu H, Rangel RH, Lavernia EJ (1995) Modeling of molten droplet impingement on a non-flat surface. Acta Metall Mater 43(5):2053–2072
91. Lee AK, Sanchez-Caldera LE, Chun JH, Suh NP (1989). In: McCandish LE (ed) Multicomponent ultrafine microstructures, MRS conference series. Metals Research Society, Pittsburgh, PA, vol 1342, p 87
92. Liu H, Zeng ZX, Lavernia EJ (1993) Processing maps for reactive atomization and development processing. Scr Metall Mater 29:1341–1344
93. Majagi SI, Ranganathan K, Lawley A, Apelian D (1992) In-Situ spray forming of metal matrix composites, in Proceedings of IRC 92: Processing. In: Loretto MH, Beevers CJ (eds) Properties and Application of Metallic and Ceramic Materials, vol 1. MCE Publishers Ltd, Birmingham, UK, pp 135–140
94. Mutasim ZZ, Smith RW (1992). In: Bernecki TF (ed) Thermal spray coatings: properties, processes and applications. ASM International, Materials Park, Ohio, USA, p 273
95. Tiwari R, Boulos MI, Jiang XL, Gitzhofer F (1994) Reactive plasma spray of refractory composites, presented at thermal spray synthesis of composites, 1994 materials week. ASM International, Materials Park, Ohio, USA, Chicago, Illinois
96. Castro RG (1994) Molybdenum disilicide composites produced by plasma spray forming, a presentation at the symposium on thermal spray synthesis of composites, materials week, Chicago, Illinois, USA
97. Lawrynowicz DE, Wolfenstine J, Nutt SR, Hurt AM, Lavernia EJ (1995) Reactive synthesis and characterization of $MoSi_2$/SiC using low pressure plasma deposition and methane. Scr Metall Mater 32(5):689–693
98. Smith RW (1993) Powder Metall Int 25:9
99. Kacar AS, Rana F, DM Stefanescu (1991) Mat Sci Eng A 135:95
100. Rohatgi PK, Asthana R, Yadav RN, Ray S (1990) Metall Trans A 21A:2073
101. Ayers JD (1984) Wear 97:249
102. Ayers JD, Schaefer RJ, Robey WP (1981) J Metals 33(8):19
103. Ayers JD, Bolster RN (1984) Wear 93:193
104. Rohatgi PK, Asthana R, Das S (1986) Int Metals Rev 31:115
105. Tiwari R, Herman H (1991) Scripts Metallurgica Materialia 25:1103

106. Gupta M, Mohamed FA, Lavernia EJ (1991) Heat transfer mechanisms and their effects on microstructure during spray atomization and co-deposition of metal matrix composites. Mater Sci Eng, A 144(1–2):99–110
107. Perez RJ, Zhang J, Gungor MN, Lavernia EJ (1993) Damping behavior of 6061 Al/graphite MMCs. Metall Trans A 24A:701–712
108. Maher PP, Grant PS, Cantor B, Katerman L (1990) Manufacture of spray formed aluminum-based alloys and composites, ICSF 90, p 31
109. Lee DJ, Vaudin MD, Handwerker CA, Katter UR (1988) Mat Res Symp Proc 120:293
110. Lo SH, Dionne S, Carpenter G, Zimcik D (1989). In: Lin RY, Arsenault RJ, Martin GP, Fishman SG (eds) Interfaces in metal matrix composites. The Minerals, Metals and Materials Society, Warrendale, PA, p 1675
111. Carpenter GJC, Lo SH (1992) J Mat Sci 27:1827
112. Lin Yaojun, Zhou Yizhang, Lavernia Enrique J (2004) Microstructural characterization of oxides in 5083 Al synthesized by reactive atomization and deposition. J Mater Res 19 (10):3090–3099
113. Lin Y, Zhou Y, Lavernia EJ (2004) An analytical model for the oxide size in Al alloys synthesized by reactive atomization and deposition. Metall Mat Trans 35A:3265–3275

Magnetically Induced Cavitation for the Dispersion of Particles in Liquid Metals

M. Sarma, I. Grants, A. Bojarevics and G. Gerbeth

Abstract A contactless excitation of cavitation is possible by superposition of induction heating with a static axial magnetic field. This creates an alternating electromagnetic body force in a liquid metal which in turn produces pressure oscillations. Using this method, the onset of cavitation has been clearly observed in various liquid metals (tin, zinc, aluminum, steel SAE 304) at pressure oscillations in the range of 28…50 kPa. The present study aims to extend the previous work by producing steel metal matrix composites (MMC) and assessing the feasibility of the proposed method for particle dispersion in steel. Stainless steel (SAE 316L) samples with different ceramic inclusions, e.g. TiN, Al_2O_3, TiB_2 as well as others, have been created. It has been demonstrated that the cavitation onset in the liquid steel varies extensively and depends on the cavitation nuclei rather than the strength of acoustic pressure. The microstructure of the produced samples has been analyzed using SEM and EDS.

Keywords MMC production · Steel composites · Cavitation treatment

Introduction

With addition of ceramic particles to a metal matrix it is possible to tailor the base material for specific needs, for instance, increase the yield and tensile strength while maintaining a good ductility. Mostly aluminum (Al) [1–4], magnesium (Mg) [5–8] and titanium (Ti) [9, 10] and their alloys are produced as composite materials and applied practically with the inclusions ranging from different sized particles to fibers. Recent developments with light metal alloys show the vast potential of

M. Sarma (✉) · I. Grants · G. Gerbeth
Institute of Fluid Dynamics, Helmholtz-Zentrum Dresden-Rossendorf, Bautzner Landstraße 400, 01328 Dresden, Germany
e-mail: m.sarma@hzdr.de

I. Grants · A. Bojarevics
Institute of Physics, University of Latvia, Salaspils, Latvia

T. S. Srivatsan et al. (eds.), *Metal-Matrix Composites Innovations, Advances and Applications*, The Minerals, Metals & Materials Series,
https://doi.org/10.1007/978-3-319-72853-7_12

composites, for example, addition of 14 vol% SiC nanoparticles inside a Mg matrix enhanced the strength, stiffness and plasticity which in turn allowed reaching very high specific yield strength (710 MPa) and specific modulus (86 GPa) [5]. However, research with steel composites is more limited due to higher working temperatures and harsher working environment which the particles must sustain. Still, one of the possible nanocomposites is the oxide dispersion strengthened (ODS) steel which has been proven of great value in nuclear fusion and fission systems because of higher tolerance against radiation, higher operational temperature and creep resistance than standard steels [11, 12]. A relatively novel material group, which is required by automotive as well as aerospace industries, is high modulus steels (HMS) which aim to reduce the weight while increasing the stiffness of the materials [13, 14].

The main MMC production routes are powder metallurgy or casting. In the simplest case, the ceramic particles are mixed together with the metal matrix powder, cold pressed and sintered, with plastic working, e.g. extrusion, being the last step. Before that an additional step of mechanical alloying might be taken to create a composite powder [9, 10]. The main drawbacks of this method are the high costs, long processing time and the limited usage of the end product, which in most cases are simple shapes with small dimensions and not usable for structural applications. In the liquid metal route, the particles are inserted directly in the melt which is stirred to distribute them [1, 8]. However, the reinforcing element should be chemically stable inside the melt as well as to have a good wetting behavior. A composite material can also be made by infiltration where a light metal is squeezed inside a porous, ceramic preform [2, 6, 7].

A rather simple solution for dispersing the particles in a liquid metal is to apply ultrasonic treatment—cavitation created by ultrasound can break particle clusters, clean the particle surface and support the dispersion [15, 16]. Suslick et al. [17] showed that the collapse of bubbles can create "hot spots" with effective temperatures of 5000 K, pressures of 1000 atm, and heating and cooling rates above 10^{10} K/s. Different methods which incorporate cavitation have been proposed and tested with various aluminum and magnesium alloys [3, 4, 8]. However, due to the much higher melting temperature of steel direct introduction of the ultrasonic probe into the melt is not possible. The interface between the probe and the metal is exposed to harsh working conditions which lead to a fast degradation of the probe as well as contamination of the melt. Therefore, for industrial applications a contactless ultrasonic treatment of the melt is needed which can be achieved by electromagnetic vibration—a method first proposed by Vives [18].

A technique that could be used to create steel nanocomposites by a casting route was proposed by Grants et al. [19]. There a steady 0.5 T strong axial magnetic field is applied to a liquid metal zone heated by electromagnetic induction. Superimposition of both fields creates an alternating radial magnetic body force in a molten sample which results in an acoustic pressure field that generates cavitation. Initially, the experiments were limited to light metals, i.e. tin, zinc, aluminum. The first investigation to admix different ceramic particles in stainless steel is described in [20]. While the concept of contactless melt vibration to create cavitation was shown to be promising, the particle admixing was insufficient. The current study aims to improve the admixing by using higher static magnetic field, longer processing time as well as improved preparation of the particles.

Experimental Details

Different ceramic particles (approx. wt% = 0.2) were mixed together with a stainless-steel powder (Cr16-18, Ni10-14, Mo2-3, C0.03, 5…20 μm) and pressed (p = 12 kN in 1120 °C) in a tablet form (d = 20 mm, h = 5 mm). The characteristics are shown in the Table 1. The experimental set-up can be seen in Fig. 1. The tablet (1) is placed between two cooled steel supports (2) that limit the axial oscillations of the molten part and are melted together with the tablet. The melting is done in an argon atmosphere with variable pressure between 1… 600 mbar in an enclosed quartz tube (3) and water-cooled copper caps (4). An alternating current (AC) induction coil (5) is used for melting the metal and a direct current (DC) magnetic field is directed by the magnet poles (6). The maximum field strength is 0.8 T. The maximum applied r.m.s. current is I_{AC} = 1300 A which corresponds to B_{AC} = 0.130 T, with frequency of the AC source being 14 kHz. The cavitation signal is measured using 4 piezo elements. After the melting, a cylindrical sample with approximate dimensions of h = 56 mm (melt zone of about 25 mm), d = 20 mm is obtained which then is cut for examination. The melting was done for approximately 3 h, while the melt vibration for 2 h.

Table 1 Sizes of the used ceramic particles

Particle	Al_2O_3	ZrO_2	TiC	TiN	CeO_2	TiB_2
Size	20 nm	800 nm	800 nm	800 nm	<5 μm	2–6 μm

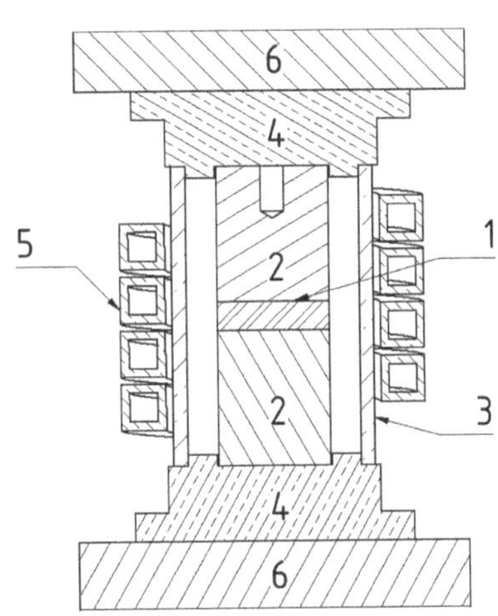

Fig. 1 Scheme of the experimental set-up: 1-sample, 2-sample holders, 3-quartz glass, 4-water cooled copper supports, 5-induction coil, 6-electromagnet pole shoes

The superposition of the AC and DC magnetic fields creates pressure oscillations in the sample. For the used frequency, and assuming a simplified case (see [19] for extended solution), the pressure amplitude p_A of those oscillations is given by

$$p_A = \frac{B_{AC} B_{DC}}{\mu_0} \tag{1}$$

where μ_0 is the vacuum magnetic permeability. In the current experiment, the maximum pressure amplitude is around 80 kPa. Vives [18] estimated that for aluminum the minimum cavitation threshold lies in range of $p_A = 65\ldots80$ kPa, whereas Abramov [21] estimated a broad range of 350…850 kPa for different metals. However, in the mentioned experiments the pressure is a sum of the static and electromagnetic pressure. Only when the second term is higher than the first, the negative pressure generates a tension in the liquid, which enables the production of a cavitation. The previous experiments with steel showed that the maximum achievable negative pressure amplitude of $p_A = 50$ kPa provides mixed results—while in some cases a full-scale transient cavitation signal was achieved, in others cavitation was not obtainable. Hence, the occurrence of cavitation cannot be defined by a precise threshold as it depends on other factors, i.e. amount of nuclei sites—impurities and dissolved gas in the melt.

Experimental Results and Discussion

Observation of the Cavitation

The acoustic spectrum of a cavitation is characterized by sub-harmonics of the drive frequency $f_n = f_0/n$, $n = 2, 3\ldots$ as well as ultra-harmonics $mf_0 \pm f_n$, $m = 1, 2\ldots$ emitted as a sound signal by the collapsing bubbles [22, 23]. To be sure that cavitation conditions are reached, this signal is recorded using the piezo elements. The recorded sound spectra of different samples are shown in Fig. 2. Application of DC field can be recognized with the broadening of the recorded signal. In all experimental trials, the pressure in the vessel was varied by periodically adding and removing argon to promote gas diffusion inside the melt, thus increasing the gas fraction and cavitating bubble entrapment. However, as previous experiments had shown, then the uncertainty of the cavitation remains as different samples experience different strength of the cavitation. A, presumably, full scale was reached only with one sample—ZrO_2 (b) where a slight broadband noise can be observed around the 100th min mark. In the case of TiN (d) very weak sub-harmonic can be observed starting from the 40th min and being present until the 100th min. Melts with CeO_2 (e) and TiB_2 (f) follow a similar pattern—the cavitation signal can be observed immediately when the DC field is applied (33rd min and 17th min accordingly), however, the phenomena is highly intermittent, i.e. having many interruptions and appearing for a short time. Al_2O_3 (a) has no sign of the 1/2 f_0

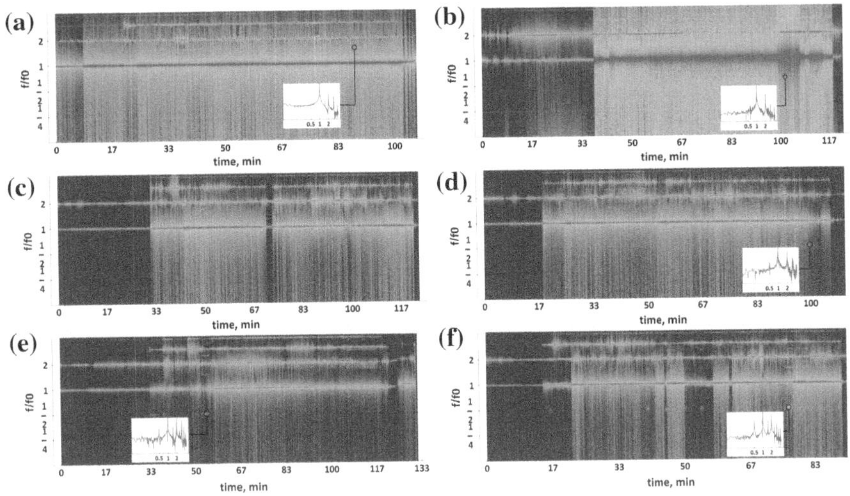

Fig. 2 Temporal dependency of the sound power spectrum when cavitation is applied to 6 different samples with Al_2O_3 (**a**), ZrO_2 (**b**), TiC (**c**), TiN (**d**), CeO_2 (**e**) and TiB_2 (**f**) particles. The depth of the logarithmic palette is 60 dB. Insets show a specific time spectrum where sign of a cavitation signal is marked by a circle

peak, however, there is a stable 2/3 f_0 signal which can be considered as a sign of cavitation. There was no record of cavitation being present in the case of TiC (c). Comparing the current results with previously obtained [20] ones, where tablets had higher particle concentration, but also much higher porosity while processed with weaker DC field, it is possible to suggest that the cavitation threshold is much more influenced by the amount of nucleus sites as well as the gas content rather than the strength of the processing parameters. Additionally, forced gas diffusion doesn't give the expected result if the initial gas fraction is low. Nevertheless, one can still expect a direct influence of this electromagnetic vibration on the overall melt mixing, structure and particle distribution.

Dispersion of Particles Using Cavitation

The samples with the particle inclusions were examined with an optical microscope and scanning electron microscope (SEM) (Zeiss EVO 50) with energy dispersive X-ray spectroscopy (EDS). Titanium compounds experience similar effects as seen in Fig. 3—particle agglomerates in a wide size range—less than 1 μm inside the steel matrix (TiC and TiB in (c) and (d) accordingly) and up to tens of μm in the outer shell (TiC and TiB in (a) and (b) accordingly)—can be found inside the steel matrix in a eutectic form. A small amount of the particles is pushed out of the matrix which then forms a thin layer on the edge of the samples as shown in Fig. 4

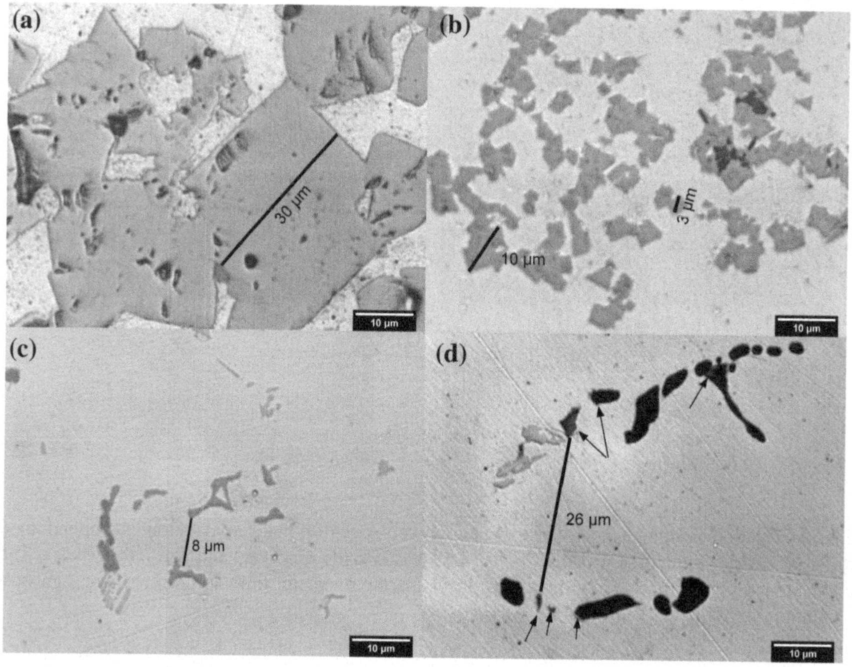

Fig. 3 Overview of TiC (left) and TiB_2 (right) microstructure. Micrographs reveal that the inclusion sizes vary between edge (**a**, **b**) and matrix (**c**, **d**). TiC (**a**) experiences much larger structures on the edge of the sample, however, the difference diminishes in the matrix. Arrows in (**d**) indicate the TiB_2 phases which can be found both attached to MgS phase (colored black) or separately in the matrix (dark gray)

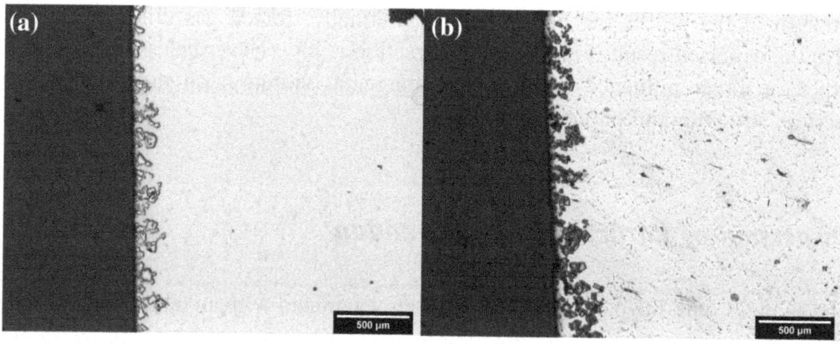

Fig. 4 TiN (**a**) and TiC (**b**) layered structure near the edge of the sample (steel matrix light grey)

(a) for TiN and (b) for TiC. It can be seen from Fig. 3 that the inclusions are not dispersed evenly and have a broad range of interspatial length. Ti phase can be found either separately in the matrix or attached to MgS phase which is also confirmed by EDS analysis of the TiN sample in Fig. 5. The noticeable difference

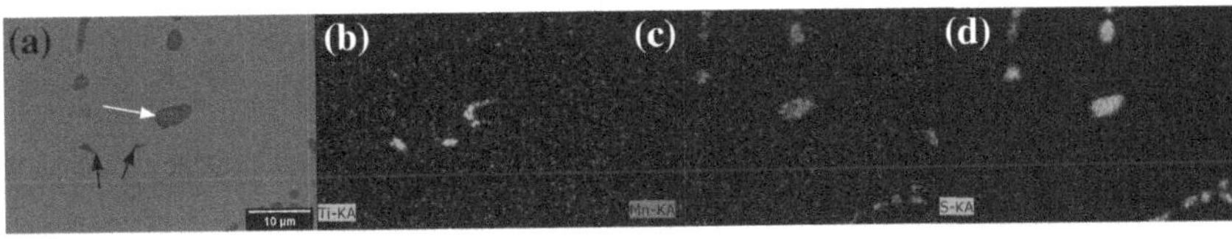

Fig. 5 EDS analysis of a TiN inclusions (**a**) shows that TiN (**b**) compounds in the matrix can be found either separately [denoted with black arrows in (**a**)] or together with MgS (**c**, **d**) phase (white arrow)

between the particle sizes in Fig. 3 (a) and (b), can be explained by the influence of cavitation as it was not present in the TiC sample. However, more experimental data is needed for a statistical validation.

Tablets with oxide particles did not melt fully with the unmelted part being around 10–30% of the initial mass. Very few, small clusters of ZrO_2 can be found near the edge of the sample as shown in Fig. 6 (b), however there are no clear indications of particles in the matrix. Figure 6 (a) shows the region of ZrO_2 sample around 1 mm from the edge where it is possible to distinguish cavities along grain boundaries. The explanation of such phenomena and the missing particles could be that ZrO_2 agglomerated in this region and was polished out because of a weak encapturing by the steel matrix. However, if the particles where present then there is uncertainty of why such a structure is formed only in few places and not in the middle nor at the edge of the sample. The SEM image in Fig. 7 (a–c) show a few CeO_2 particle agglomerate inside the metal matrix which is attached to the MgS phase. Similar structures can be found throughout the matrix, however with irregular distances. Additionally, there are many cavities inside the matrix with different shapes as seen in Fig. 7 (d) as bright spots. While the smallest, because of being spherical, can be related to gas inclusions and provide hints of the nuclei size for production of cavitation, the larger ones have particle like shapes. Hence, it could be assumed that some amount of particles were removed in the polishing process, but further investigation is needed to support this claim. Lastly, Fig. 8

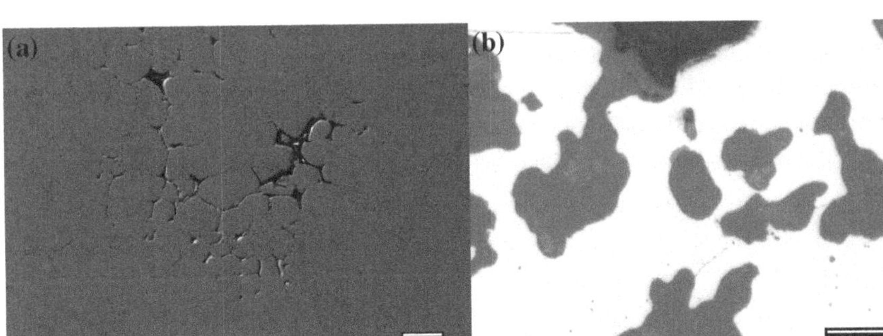

Fig. 6 Cavity formation on the grain boundaries when melted together with ZrO_2 (**a**) and ZrO_2 agglomerations near the edge of the sample (**b**)

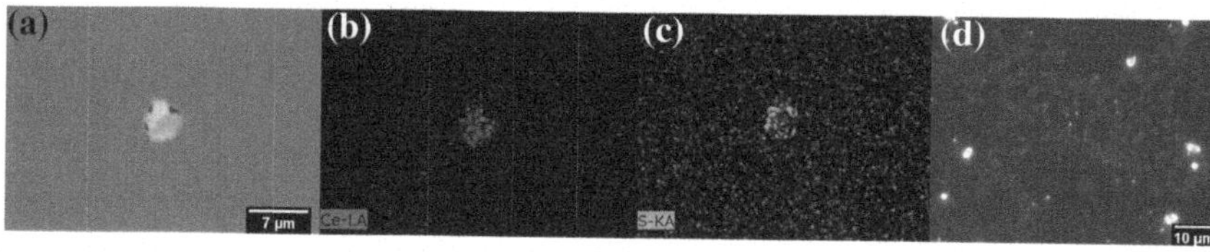

Fig. 7 EDS analysis of a CeO_2 particle cluster (**a**) showing that Ce phase (**b**) is forming in clusters with MgS (**c**) phase. Micrograph (**d**) allows to see the amount of cavities (bright phase) inside the matrix

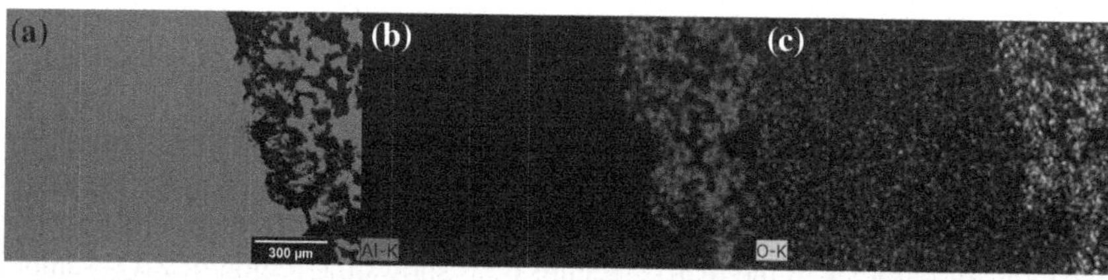

Fig. 8 EDS analysis of Al_2O_3 cluster near the edge (part of an unmelted tablet)

shows an EDS analysis of Al_2O_3 cluster which is part of an unmelted tablet. However, besides a few small inclusions attached to the edge of the sample, no other agglomerates could be found and it was not possible to reach the necessary resolution with the used SEM to distinguish a single particle hence it is unknown if the rest of Al_2O_3 are in the matrix.

There is no clear explanation why oxide sample tablets did not fully melt nor it is clear where the oxide particles are agglomerated. While it is difficult to estimate how much of the particles were left in the unmelted part, it is only a fraction, thus it is important to investigate these samples in greater detail. All Ti compounds are in eutectic form inside the matrix which is similar to what is found when particles are formed in situ in Fe-Ti-B melt [24]. However, this also means that Ti particles cannot be considered stable inside the melt as they react with liquid steel and lose their initial form, i.e. particles do not survive the processing. Comparison of TiB_2, CeO_2 and TiN with the same powders in the previous experiment show that the used compounds are wetted better if they are exposed for longer period to the electromagnetic vibration in the liquid steel as the mentioned compounds have been found inside the matrix and not agglomerated near the edges or completely pushed out of the matrix. Still, more experiments are necessary to validate this claim.

Conclusions

Contactless induction of cavitation in liquid steel for different particle dispersion has been investigated. It is not yet possible to assess the feasibility of this method as the particle dispersion is a highly complex problem which is dependent not only on the method used and the strength of applied forces, but also on the physical and chemical compatibility between the selected particles and metal matrix. The current experiments have shown that successful production of cavitation is much more dependent on the number of nuclei sites inside the volume rather than on the exerted force on the melt. Ti compounds show similar characteristics by transforming to eutectic form when exposed to the liquid steel environment for extended periods as well as similar dispersion properties. If ZrO_2 has been removed from the matrix by polishing then it can be assumed that this oxide might have much weaker bonding with the steel matrix as well as possibly weaker wetting while CeO_2 is mostly attached to MgS and thus has improved wetting.

Acknowledgements The financial support of the Helmholtz Alliance "Liquid Metal Technologies—LIMTECH" is gratefully acknowledged.

References

1. Tahamtan S, Emamy M, Halvaee A (2014) Effects of reinforcing particle size and interface bonding strength on tensile properties and fracture behavior of Al-A206/alumina micro/nanocomposites. J Compos Mater 48(27):3331–3346. https://doi.org/10.1177/0021998313509860
2. Sharifi H, Nasresfahani MR (2016) Investigation into the kinetic behavior of molten aluminum pressureless infiltration into SiC preforms. Int J Mater Res 107(10):954–959. https://doi.org/10.3139/146.111422
3. Lü SL, Xiao P, Wu SS, Fang XG (2017) Preparation and squeeze casting of nano-SiCP/A356 composite assisted with ultrasonic vibration process. Mater Sci Forum 879:1188–1193. doi: https://doi.org/10.4028/www.scientific.net/MSF.879.1188
4. Yan H, Huang Z-X, Qiu H-X (2017) Microstructure and mechanical properties of CNTs/A356 nanocomposites fabricated by high-intensity ultrasonic processing. Metall Mater Trans A 48(2):910–918. https://doi.org/10.1007/s11661-016-3872-1
5. Chen L-Y, Xu J-Q, Choi H, Pozuelo M, Ma X, Bhowmick S, Yang J-M, Mathaudhu S, Li X-C (2015) Processing and properties of magnesium containing a dense uniform dispersion of nanoparticles. Nature 528(7583):539–543. https://doi.org/10.1038/nature16445
6. Braszczyńska-Malik KN, Kamieniak J (2017) AZ91 magnesium matrix foam composites with fly ash cenospheres fabricated by negative pressure infiltration technique. Mater Charact 128:209–216. https://doi.org/10.1016/j.matchar.2017.04.005
7. Zhang X, Zhang Q, Hu H (2014) Tensile behaviour and microstructure of magnesium AM60-based hybrid composite containing Al_2O_3 fibres and particles. Mater Sci Eng A 607:269–276. https://doi.org/10.1016/j.msea.2014.03.069
8. Wang XJ, Wang NZ, Wang LY, Hu XS, Wu K, Wang YQ, Huang YD (2014) Processing, microstructure and mechanical properties of micro-SiC particles reinforced magnesium matrix composites fabricated by stir casting assisted by ultrasonic treatment processing. Mater Des 57:638–645. https://doi.org/10.1016/j.matdes.2014.01.022
9. Li S, Sun B, Imai H, Mimoto T, Kondoh K (2013) Powder metallurgy titanium metal matrix composites reinforced with carbon nanotubes and graphite. Compos A Appl Sci Manuf 48 (1):57–66. https://doi.org/10.1016/j.compositesa.2012.12.005
10. Munir KS, Zheng Y, Zhang D, Lin J, Li Y, Wen C (2017) Improving the strengthening efficiency of carbon nanotubes in titanium metal matrix composites. Mater Sci Eng A 696:10–25. https://doi.org/10.1016/j.msea.2017.04.026

11. Dubuisson P, Carlan YD, Garat V, Blat M (2012) ODS ferritic/martensitic alloys for sodium fast reactor fuel pin cladding. J Nucl Mater 428(1–3):6–12. https://doi.org/10.1016/j.jnucmat.2011.10.037
12. Franke P, Heintze C, Bergner F, Weißgärber T (2010) Mechanical properties of spark plasma sintered Fe-Cr compacts strengthened by nanodispersed yttria particles. Materialpruefung/Mater Test 52(3):133–138
13. Bonnet F, Daeschler V, Petitgand G (2014) High modulus steels: new requirement of automotive market. How to take up challenge? Can Metall Q 53(3):243–252. https://doi.org/10.1179/1879139514Y.0000000144
14. Springer H, Baron C, Szczepaniak A, Uhlenwinkel V, Raabe D (2017) Stiff, light, strong and ductile: nano-structured high modulus steel. Sci Rep 7(1) (art.no.2757). https://doi.org/10.1038/s41598-017-02861-3
15. Bittmann B, Haupert F, Schlarb AK (2009) Ultrasonic dispersion of inorganic nanoparticles in epoxy resin. Ultrason Sonochem 16(5):622–628. https://doi.org/10.1016/j.ultsonch.2009.01.006
16. Sumitomo S, Koizumi H, Uddin MA, Kato Y (2018) Comparison of dispersion behavior of agglomerated particles in liquid between ultrasonic irradiation and mechanical stirring. Ultrason Sonochem 40:822–831. https://doi.org/10.1016/j.ultsonch.2017.08.023
17. Suslick KS, Didenko Y, Fang MM, Hyeon T, Kolbeck KJ, McNamara III, WB, Mdleleni MM, Wong M (1999) Acoustic cavitation and its chemical consequences. Philos Trans R Soc A: Math Phys Eng Sci 357(1751):335–353
18. Vivès C (1996) Crystallization of aluminium alloys in the presence of cavitation phenomena induced by a vibrating electromagnetic pressure. J Cryst Growth 158(1–2):118–127
19. Grants I, Gerbeth G, Bojarevičs A (2015) Contactless magnetic excitation of acoustic cavitation in liquid metals. J Appl Phys 117(20) (art.no.204901). https://doi.org/10.1063/1.4921164
20. Sarma M, Grants I, Kaldre I, Bojarevics A, Gerbeth G (2017) Casting technology for ODS steels—dispersion of nanoparticles in liquid metals. IOP Conf Series: Mater Sci Eng 228(1):012020. https://doi.org/10.1088/1757-899X/228/1/012020
21. Abramov OV (1987) Action of high intensity ultrasound on solidifying metal. Ultrasonics 25(2):73–82. https://doi.org/10.1016/0041-624X(87)90063-1
22. Neppiras EA (1980) Acoustic cavitation. Phys Rep 61(3):159–251. https://doi.org/10.1016/0370-1573(80)90115-5
23. Cramer E, Lauterborn W (1982) Acoustic cavitation noise spectra. Appl Sci Res 38(1):209–214. https://doi.org/10.1007/BF00385950
24. Aparicio-Fernández R, Springer H, Szczepaniak A, Zhang H, Raabe D (2016) In-situ metal matrix composite steels: effect of alloying and annealing on morphology, structure and mechanical properties of TiB_2 particle containing high modulus steels. Acta Mater 107:38–48. https://doi.org/10.1016/j.actamat.2016.01.048

An Engineered Magnesium Alloy Nanocomposite: Mechanisms Governing Microstructural Development and Mechanical Properties

Sravya Tekumalla, Shikhar Bharadwaj, T. S. Srivatsan and Manoj Gupta

Abstract A magnesium alloy-based nanocomposite [Mg-1.8Y-0.4Ce/1.53ZnO] was developed using the technique of disintegrated melt deposition (DMD) followed by hot extrusion. In this paper, an attempt is made to systematically present and elegantly discuss the conjoint influence of yttrium, cerium and zinc oxide additions to pure magnesium. The developed nanocomposite revealed a combination of superior microstructure and mechanical properties (microhardness, strength and ductility). The microscopic mechanisms governing both strength and ductility of the engineered nanocomposite were examined through carefully conducted study of microstructure using an optical microscope, scanning electron microscope, and x-ray diffraction (XRD) texture analysis. Microstructure-mechanical property correlation studies were performed with the prime objective of understanding and rationalizing the microscopic mechanisms governing mechanical behavior.

Keywords Magnesium alloy nanocomposite · Extrusion · Microstructure, strength · Ductility

Introduction

Magnesium, owing to its excellent strength-to-weight [σ/ρ] ratio, is a viable candidate for both selection and use in a spectrum of weight-critical applications in the industries spanning automotive, other ground transportation and air transportation

S. Tekumalla · S. Bharadwaj · M. Gupta (✉)
Department of Mechanical Engineering, National University of Singapore, Singapore 117 576, Singapore
e-mail: mpegm@nus.edu.sg

T. S. Srivatsan
The University of Akron, Akron, OH 44325-3903, USA

T. S. Srivatsan et al. (eds.), *Metal-Matrix Composites Innovations, Advances and Applications*, The Minerals, Metals & Materials Series,
https://doi.org/10.1007/978-3-319-72853-7_13

while concurrently finding use for even commercial products. Due to its limited ductility, at room temperature, coupled with highly anisotropic nature, magnesium and its alloy counterparts have in recent years been the subject of several intensive studies undertaken with the prime objective of overcoming the limitations in its properties. A viable and economically affordable approach of overcoming the limitation arising from low ductility is by the prudent selection and use of alloying elements, like elements belonging to the rare-earth family. These elements contribute in an observable way to weakening of the texture and in the process, contribute to improving the ductility of both magnesium and its alloys [1, 2]. The elements yttrium (Y) and cerium (Ce) have shown a promising trend in improving the ductility of cast magnesium as well as wrought magnesium [3, 4] with a concomitant drop in the yield strength. It has been reported that the addition of more than 0.5 wt% Ce resulted in excessive oxidation of the surface during extrusion [5]. Further, the increasing trend in the cost of rare earth elements may not be appropriate for the magnesium alloys containing rare earth elements. This is especially true for those magnesium alloys that contain a high concentration of the rare earth elements. Hence, yttrium was chosen and added since it behaves like rare earth element for concentration less than 2 pct. Besides, cerium is also added to magnesium in minor quantities (0.4%) in the present study. A new approach has been being tried in the present study to offset the high cost of the rare earth elements by reinforcing the magnesium alloys rare earth elements with nano particulates to form or engineer nanocomposites. Accordingly, the present study focusses on the development and characterization of Mg-1.8Y-0.4Ce/1.53ZnO nanocomposite. Emphasis is placed upon studying the effect of addition of 1.8 yttrium (Y), 0.4 cerium (Ce) and nano zinc-oxide (ZnO) particles to magnesium.

Experimental Procedures

Synthesis

Magnesium turnings of 99.9% purity [supplied by Acros Organics, NJ, USA] and Mg-30%Y and Mg-30%Ce master alloy of purity 99% [supplied by Sunrelier Metal Co. Limited, Pudong, Shanghai, China], Zinc oxide (ZnO) nano powder of size 90–200 nm having a purity of 99.995% were used for the synthesis of Mg-1.8Y-0.4Ce/1.53ZnO nanocomposite. The materials [pure Mg and Mg-1.8Y-0.4Ce/1.53ZnO nanocomposite] were synthesized using the liquid metallurgy route technique of Disintegrated Melt Deposition (DMD) [6]. The resultant ingot measured 40 mm in diameter and was precision machined to get an ingot that was 36 mm in diameter. The machined ingot was now soaked at 450 °C for 2 h. Hot extrusion was then carried out using a 150 T hydraulic press and an extrusion ratio of 20.25:1 to obtain rods that measured 8 mm diameter.

Characterization of Physical Properties: (i) Density, (ii) Porosity, and (iii) Coefficient of Thermal Expansion

The density and porosity measurements were carried out on polished samples having an aspect ratio of 1 using the Archimedes' Principle [7]. The liquid chosen for purpose of immersion was distilled water. The readings were taken for 8 samples using an electronic balance [Model: A&D HM-202] with an accuracy of ±0.1 mg.

A thermo-mechanical analyser [Model: INSEIS TMA PT 1000LT] was used to obtain the coefficient of thermal expansion (CTE) of the as-extruded pure magnesium and the engineered Mg-1.8Y-0.4Ce/1.53ZnO nanocomposite. A heating rate of 5 K/min was maintained along with an argon gas flow rate of 100 cm^3/min with the prime objective of preventing oxidation. Displacement of the test samples, under a constant load of 100 mN, was measured as a function of temperature (50–400 °C) using an alumina probe. These measurements were used to determine the Coefficient of thermal expansion (CTE) of the samples.

Characterization of Initial Microstructure

Samples of pure magnesium and the engineered Mg-1.8Y-0.4Ce/1.53ZnO nanocomposite were subjected to standard metallographic preparation technique, which involved grinding, polishing and etching and subsequent examination in an optical microscope for purpose of characterizing the microstructure. An optical microscope [Model: Leica DM2500 M] was used to obtain micrographs, over a range of allowable magnifications, to help reveal both the size and morphology of grains. A Scion image analysis software was used to compute the grain size. Both a scanning electron microscope [Model: JEOL JSM-5600LV SEM] and a field emission scanning electron microscope [Model: Hitachi FESEM-S4300] coupled with Energy Dispersive Spectroscopy (EDS) were used for studying the composition and distribution of the matrix strengthening precipitates and reinforcing secondary-phase particles.

Characterization of Mechanical Properties: (i) Micro-Hardness, and (ii) Compressive Properties

Micro-hardness measurements were made using a Matsuzawa MXT 50 automatic digital micro-hardness tester using a test load of 245.2 mN and a dwell time of 15 s. The micro-hardness test was conducted in accordance with specifications detailed in ASTM standard E384-11e1.

The uniaxial compression tests on test samples of both as-extruded pure magnesium and engineered Mg-1.8Y-0.4Ce/1.53ZnO nanocomposite was conducted using a MTS 810 test machine in accordance with specifications detailed in ASTM test method E9-09. The test samples were deformed at a strain rate: 0.01 $\min^{-1}$. For each composition, a minimum of 5 tests was conducted to obtain reproducible results.

Results and Discussion

The synthesis of pure magnesium and the engineered Mg-1.8Y-0.4Ce/1.53ZnO nanocomposite was successfully done and a visual observation of the extruded rods did not reveal the presence of observable defects.

Microstructure

The optical micrographs (Fig. 1a, b) reveal the average grain size of the two materials that are neatly summarized in Table 1. There is an observable decrease in average grain size of the Mg-1.8Y-0.4Ce/1.53ZnO nanocomposite when compared one-on-one with that of pure magnesium by as much as 78%, indicating the occurrence of accelerated recrystallization mechanism [8]. Since the temperature of extrusion is 450 °C and the solubility of cerium in magnesium at that temperature is 0.06 wt% [9], the presence of intermetallic particles in the microstructure is inevitable. During hot extrusion of the alloy, these intermetallic particles contribute to pinning of the grain boundaries that does contribute in a small way to refining the grain size. The reduction in grain size of the hot extruded Mg-1.8Y-0.4Ce/1.53ZnO

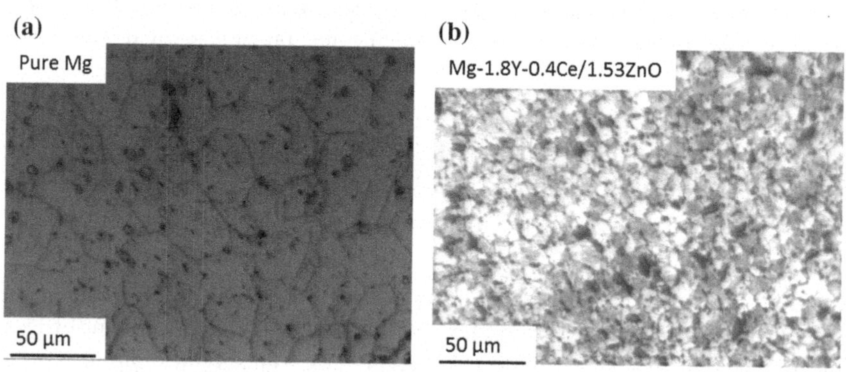

Fig. 1 Optical micrographs showing the matrix grain characteristics of: **a** Pure Mg, and **b** Mg-1.8Y-0.4Ce/1.53ZnO nanocomposite

Table 1 A summary of the grain size of the two materials

Composition	Grain Size (μm)
Pure magnesium	26.8 ± 4.5
Mg-1.8Y-0.4Ce/1.53ZnO	5.9 ± 1.1 (−78%)

'(±x%)' indicate the increase/decrease in the property with respect to pure magnesium by x%

nanocomposite can be attributed to the fact that during extrusion, both the zinc oxide (ZnO) nano particles and the (Mg, Ce) intermetallic particles tend to restrict grain growth by acting as barriers for the easy migration of the grain boundaries.

The scanning electron micrograph (SEM) of the Mg-1.8Y-0.4Ce/1.53ZnO nanocomposite (Fig. 2a) indicates a uniform distribution of intermetallic particles of (i) Mg–Ce, and (ii) Mg–Y in the size range of about 1–2 μm. In the engineered nanocomposite, there is a near uniform distribution of ZnO particles in the matrix with an evidence of traces of agglomeration as has been reported in the published

(a)

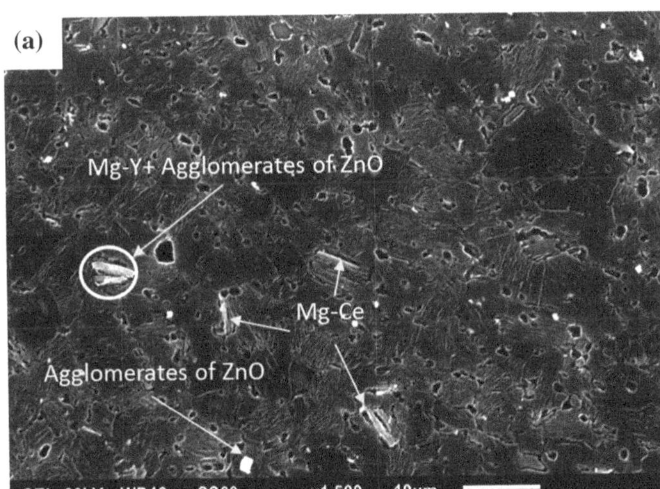

(b)

Elem...	Weight%	Atomic%	Compd%	Formula
O K	8.93	14.10	12.28	O2C
Mg K	70.65	73.41	0.00	
Zn K	5.81	2.25	0.00	
Y L	11.25	3.20	0.00	
Ce L	0.00	0.00	0.00	
C	3.35	7.05		
Totals	100.00			

Fig. 2 **a** Scanning electron micrograph showing the distribution of Mg–Ce, Mg–Y intermetallic particles in the matrix along with ZnO particles (**b**) EDS elemental quantification of the inset shown in (**a**)

literature [10]. The Mg–Ce and Mg–Y particles exhibit two different kinds of morphologies; one being rod shaped, and the other being blocky phase. From the EDS elemental composition summarized in Fig. 2b, it is evident that the zinc oxide (ZnO) particles are bonded to the Mg–Y particles. Furthermore, agglomeration of the particles was evident as seen in the micrograph shown in Fig. 2a.

Density and Porosity

The results of the density measurements conducted on the extruded samples of pure magnesium and the engineered Mg-1.8Y-0.4Ce/1.53ZnO nanocomposite are presented in Table 2. From the results contained in this table, dense materials were synthesized. Calculation of volumetric porosity reveals the engineered nanocomposite Mg-1.8Y-0.4Ce/1.53ZnO to have a low porosity level. The results of coefficient of thermal expansion (CTE) (Table 2) indicate lower value of CTE for the nanocomposite relative to pure Mg thereby providing an indication for higher dimensional stability.

X-Ray Diffraction Studies

The Bragg angles, intensity peaks and interplanar spacing (d) obtained were matched with the standard values for magnesium, cerium, yttrium, and zinc oxide related phases. Secondary phases like $Mg_{12}Ce$ were not easily detected due to the presence of small amounts of cerium. The XRD results for pure magnesium are also included for purpose of comparison. The peaks of α-Mg were the only peaks found in the alloy. Further, the intensity of the peaks at the different angles of 32°, 34° and 36° correspond to the following: (i) prismatic plane (100), (ii) basal plane (002), and (iii) pyramidal plane (101) of magnesium. It is seen from the diffractograms taken from the longitudinal section of the sample, that the peak corresponding to the basal plane had the highest intensity for pure magnesium. This indicates the presence of a strong basal texture for magnesium [11]. However, the intensity of the basal plane was not the highest for the nanocomposite. Instead, the pyramidal

Table 2 Results of physical characterization and grain size analysis

Material	Experimental density (g/cc)	Porosity (%)	Coefficient of thermal expansion (μm/K)
Magnesium	1.672 ± 0.058	0.063	25.3
Mg-1.8Y-0.4Ce/1.53ZnO	1.785 ± 0.019 (+6.7%)	0.024	24.6

'(±x%)' indicate the increase/decrease in the property with respect to pure magnesium by x pct peak had the highest intensity, indicating a gradual weakening of the texture or randomization of the texture. This does play a crucial role in governing the mechanical behavior of the ternary alloy.

Micro-Hardness

The results of micro-hardness measurements, summarized in Table 3, provide an indication of the increase in mean hardness value of the engineered Mg-1.8Y-0.4Ce/1.53ZnO nanocomposite (by ~44%) when compared to pure Mg. Based entirely on microstructural observations (Table 1 and Fig. 2), the increased strength and increased hardness of the Mg-1.8Y-0.4Ce/1.53ZnO nanocomposite can be attributed to an overall refinement in the grain size (by ~78%) in synergism with precipitation hardening effects (Fig. 3).

Table 3 Results of micro-hardness tests

Composition	Micro-hardness (Hv)
Pure magnesium	73.6 ± 5.0
Mg-1.8Y-0.4Ce/1.53ZnO	106 ± 11.3 (+44%)

'(±x%)' indicate the increase/decrease in the property with respect to pure magnesium by x%

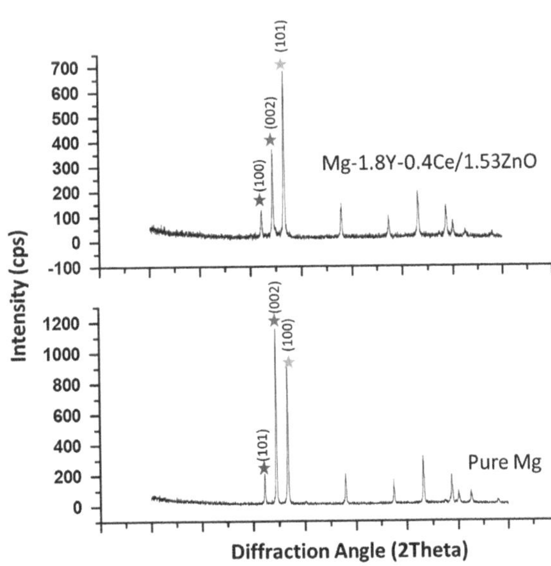

Fig. 3 X ray diffraction line profile of magnesium and Mg-1.8Y-0.4Ce/1.53ZnO nanocomposite along the extrusion direction

Response in Compression

The compressive response of the Mg-1.8Y-0.4Ce/1.53ZnO nanocomposite is compared with that of pure magnesium in Table 4. For the nanocomposite, there is an increase in yield strength (σ_{ys}) by 149 pct. and the ultimate tensile strength (σ_{UTS}) by 35 pct., with a 93 pct., increase in tensile ductility when compared to pure magnesium. Overall, the nanocomposite exhibited superior properties, in terms of both strength and ductility coupled with an increase in damage tolerance of pure magnesium by improving the energy absorption by 181%.

The increased strength and hardness of the engineered Mg-1.8Y-0.4Ce/1.53ZnO nanocomposite can be ascribed to the conjoint and mutually interactive influences of the following mechanisms.

(i) Reduced average grain size (by ~78%).
(ii) Precipitation hardening due to the presence of the different intermetallic particles.
(iii) Orowan strengthening arising from the presence of nano ZnO reinforcement.
(iv) Large difference in CTE between the reinforcement (nano particles of zinc oxide) and the matrix material (Mg-1.8Y-0.4Ce) that induces internal thermal stresses at an interface during fabrication of the composite resulting in high dislocation density, and
(v) Higher constraint on matrix deformation coupled with efficient load transfer in the composite with the addition and presence of hard, brittle and elastically deforming ZnO particles.

Thus, under the influence of an applied stress, the intermetallic particles of: (a) Mg–Y–Ce, (b) nano ZnO particles, and (c) increased grain boundary area in the matrix restrict the movement of dislocations that gradually results in dislocation pile up that contributes to increasing strength of the nanocomposite. The compressive test results also reveal a significant improvement in ductility by as much as 93% for the nanocomposite when compared with pure magnesium. The increased ductility of the engineered nanocomposite can be attributed to the reduced grain size that leads to an increase in grain boundary area as well as a change in texture as shown in the preceding section [3, 12]. The results of observation of the fracture surfaces of the

Table 4 Results of compressive tests

Composition	0.2% offset yield strength (MPa)	Ultimate compressive strength (MPa)	Ductility (%)	Energy absorbed (MJ/m^3)
Pure magnesium	62.2 ± 2	336 ± 11	15.5 ± 4.6	33.5 ± 2.8
Mg-1.8Y-0.4Ce/1.53ZnO	155 ± 12 (+149%)	454 ± 7 (+35%)	29.9 ± 0.4 (+93%)	94.3 ± 1.6 (+181%)

'(+x%)' indicates the increase in the property with respect to pure magnesium by x%

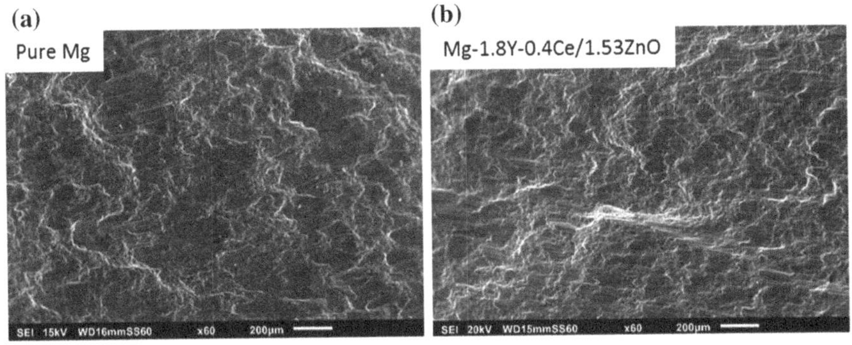

Fig. 4 Scanning electron micrographs of the fracture surfaces of test samples deformed in compression: **a** Pure magnesium, and **b** Mg-1.8Y-0.4Ce/1.53ZnO nanocomposite

deformed and failed specimens in a scanning electron microscope (Fig. 4) reveals at the macroscopic level a shear mode of failure for both pure magnesium and the nanocomposite under compression.

Conclusions

Pure Mg and Mg-1.8Y-0.4Ce/1.53ZnO nanocomposite were synthesized using the technique of disintegrated melt deposition (DMD) and subsequently processed by hot extrusion. A microstructure-property correlation was undertaken and systematically conducted, and the key conclusions are:

1. Microstructural analysis revealed a reduction in grain size by 78% for the Mg-1.8Y-0.4Ce/1.53ZnO nanocomposite when compared one-on-one with pure magnesium.
2. The nanocomposite exhibited higher hardness ($\sim$44%), yield strength ($\sim$149%), ultimate tensile strength ($\sim$35%) and reduced fracture strain by ($\sim$93%) when compared one-on-one with pure magnesium. The observed improvement in properties can be attributed to the conjoint influence of the following: (i) reduced grain size, (ii) strengthening arising from the presence and distribution of intermetallic particles, (iii) presence of hard, brittle and elastically deforming ZnO reinforcement in the metal matrix, and (iv) mismatch in coefficient of thermal expansion (CTE) between the matrix and the reinforcing phase.
3. Overall, the nanocomposite exhibited superior properties, quantified in terms of strength, ductility and damage tolerance, when compared to pure magnesium Mg. The improvement in energy absorption capability was as high as 181%.

Acknowledgements *Sravya Tekumalla* expresses gratitude to the *National University of Singapore* (NUS) for supporting her graduate study through a NUS research scholarship. The authors gratefully acknowledge the Ministry of Education Academic Research Funding for financial support [WBS# R-265-000-498-112] of this innovative and inspiring research study.

References

1. Mackenzie LWF, Pekguleryuz MO (2008) The recrystallization and texture of magnesium–zinc–cerium alloys. Scripta Mater 59:665–668
2. Bohlen J, Nürnberg MR, Senn JW, Letzig D, Agnew SR (2007) The texture and anisotropy of magnesium–zinc–rare earth alloy sheets. Acta Mater 55:2101–2112
3. Mishra RK, Gupta AK, Rao PR, Sachdev AK, Kumar AM, Luo AA (2008) Influence of cerium on the texture and ductility of magnesium extrusions. Scripta Mater 59:562–565
4. Tekumalla S, Seetharaman S, Almajid A, Gupta M (2015) Mechanical properties of magnesium-rare earth alloy systems: a review. Metals 5:1–39
5. Luo A, Wu W, Mishra R, Jin L, Sachdev A, Ding W (2010) Microstructure and mechanical properties of extruded magnesium-aluminum-cerium alloy tubes. Metall and Mat Trans A 41:2662–2674
6. Hassan SF, Gupta M (2002) Development of a novel magnesium–copper based composite with improved mechanical properties. Mater Res Bull 37:377–389
7. Ratcliffe RT (1965) The measurement of small density changes in solids. Br J Appl Phys 16:1193
8. Humphreys FJ (1979) Recrystallization mechanisms in two-phase alloys. Metal Science 13:136–145
9. Tekumalla S, Seetharaman S, Bau NQ, Wong WLE, Goh CS, Shabadi R, Gupta M (2016) Influence of cerium on the deformation and corrosion of magnesium. J Eng Mater Technol 138:031011
10. Goh CS, Wei J, Lee LC, Gupta M (2007) Properties and deformation behaviour of Mg–Y2O3 nanocomposites. Acta Mater 55:5115–5121
11. Yang W, Tekumalla S, Gupta M (2017) Cumulative effect of strength enhancer—lanthanum and ductility enhancer—cerium on mechanical response of magnesium. Metals 7:241
12. Sabat RK, Mishra RK, Sachdev AK, Suwas S (2015) The deciding role of texture on ductility in a Ce containing Mg alloy. Mater Lett 153:158–161

The Tensile Response and Fracture Behavior of a Copper-Niobium Microcomposite: Role of Surface Modification

Paul Arindam and T. S. Srivatsan

Abstract In this research study, the influence of nanocrystalline surface modification on microhardness, tensile response and fracture behaviour of an oxide dispersion strengthened copper-niobium (Cu-Nb) micro-composite was investigated. The presence of a hardened surface layer and associated compressive residual stress lead to a noticeable increase in micro-hardness and a marginal improvement in both stiffness and strength. The increase in work hardening is quantified by the monotonic stress versus strain curve. For both the as-provided and surface treated composites tensile fracture was macroscopically ductile and microscopically revealed features reminiscent of locally brittle and ductile failure mechanisms. The mechanical properties and fracture behaviour is discussed considering nature of loading and intrinsic microstructural effects.

Keywords Cu-Nb microcomposite · Surface modification · Micro-hardness Tensile response · Tensile fracture · Microstructural effects

Introduction

The ability to achieve active cooling, while the structure is still very much in service, has created an inescapable need for the chosen structural material to possess high thermal conductivity. Copper is both an ideal and attractive choice to meet this requirement. It is essential that the material, besides offering adequate strength, must also be resistant to softening and concomitant degradation upon exposure to elevated temperatures while in service. Available approach to strengthening copper is through controlled solid solution strengthening. This is normally achieved

P. Arindam
Division of Materials Science and Engineering, Department of Mechanical Engineering, The University of Akron, Akron, OH 44325-3903, USA

T. S. Srivatsan (✉)
The University of Akron, Akron, OH 44325-3903, USA
e-mail: tsrivatsan@uakron.edu

T. S. Srivatsan et al. (eds.), *Metal-Matrix Composites Innovations, Advances and Applications*, The Minerals, Metals & Materials Series,
https://doi.org/10.1007/978-3-319-72853-7_14

through carefully monitored solid solution strengthening that is made possible through a careful control by judicious alloying with other elements. However, there does occur a loss in both electrical conductivity and thermal conductivity. Introduction of a small volume fraction of a second-phase, such as inert particles, into a soft and plastically deforming copper matrix is an attractive and viable mechanism for enhancing the strength of copper while concurrently retaining both its electrical conductivity and thermal conductivity at acceptable levels. The resultant material is a dispersion strengthened (DS) alloy [1–5]. Oxides have by far been the most common second-phase particles that have successfully been used to strengthen the matrix of pure copper and its alloys [4, 5]. For the achieving optimum strengthening it is essential that the reinforcing oxide particles, essentially inert in nature, be extremely small, discontinuous and randomly dispersed through the continuous copper matrix, while concurrently having a small inter-particle spacing [1, 4]. The much superior properties of the oxide dispersion strengthened copper (DSC) material arises from the competing and mutually interactive influences of size, distribution, and interparticle spacing of the insoluble sub-microscopic aluminium oxide (Al_2O_3) particles in a high purity copper matrix. These particles range in size from 3 to 12 nm with an inter-particle spacing of 30–100 nm and a particle density of the order of 10^{16}–10^{17} cm^{-3} [2–4]. On account of their chemically inert nature, the hard, brittle and essentially elastically deforming Al_2O_3 particles exert little influence on electrical conductivity of the matrix material. Besides, the insoluble Al_2O_3 particles have the capability to retain their original size, shape, distribution and spacing even following prolonged exposure to elevated temperatures [5, 6].

Noticeable improvements in the mechanical and physical properties of dispersion strengthened copper matrix were achieved through the controlled addition of the body-centered cubic [b.c.c.] metal, such as niobium, which has limited solubility in copper [7]. The enhanced strengthening resulting from deformation processing of composites for a two-phase face-centered cubic-body centered cubic [f.c.c-b.c.c] mixture can be attributed to the presence of ribbon-like filaments of the reinforcing phase that results due to the $\langle 110 \rangle$ fibre deformation texture that develops as the two phases co-deform [8]. With time, the mechanical deformation processed copper-refractory metal (niobium) mixtures have gained adequate interest primarily because in addition to offering good electrical conductivity and thermal conductivity [9] their high ductility necessitates the need for a large degree of mechanical working, which is essential for developing adequate strength [10–12]. The presence of a fine array of the ribbon-like filaments of the reinforcing body-centered cubic (b.c.c) phase within a soft and plastically deforming face-centered cubic (f.c.c) copper matrix has been found to be both an effective and viable method for strengthening copper. On account of its limited solubility in copper, alloying through small additions of niobium will not adversely influence the electrical conductivity and thermal conductivity of copper. Also, the niobium reinforcement can retain its high strength at elevated temperatures and is an ideal material choice for use in a spectrum of elevated temperature applications [12–15].

Creating a nano-crystalline layer on the surface of a material can also contribute to noticeable improvement in mechanical properties and thereby enhance the performance. In more recent years, i.e., since the early 1990s, several mechanical surface processing techniques have been reported with the primary intent of inducing nano-crystallization on the surface. A few of these techniques are the following:

(a) Surface mechanical attrition treatment (SMAT) [16–18]
(b) Shot-peening (SP) [19]
(c) Ultrasonic shot peening (USP) [20, 21], and
(d) Laser shock peening (LSP) [22–28].

Properties of the surface to include: (i) wear resistance [29, 30], (ii) corrosion resistance [31], (iii) biocompatibility [32], and (iv) bulk mechanical properties to include tensile, fatigue and fracture, can be significantly improved through nano-crystallization of the surface [33–35].

A noticeable improvement in strength of the material is often accompanied by a loss in its ductility. For face-centered cubic (f.c.c) copper, the ductility is strongly affected by its capability to work harden, i.e., the capacity to both accumulate and store dislocations during mechanical deformation in tension. In this study, surface nano-crystallization of a dispersion strengthened copper-niobium micro-composite was carried out using the technique of ultrasonic nanocrystal surface modification. The mechanical properties (tensile) of the Cu-Nb micro-composite samples subsequent to surface nano-crystallization processing were evaluated and compared with the untreated counterpart, i.e., as-provided composite material. The conjoint influence of dispersion strengthening, niobium reinforcement and surface nano-crystallization on tensile response of the copper alloy is discussed considering the conjoint and mutually interactive influences of nature of loading (uniaxial tension), intrinsic microstructural effects, and macroscopic aspects governing fracture behaviour.

Material and Processing

The material used in this research study was an oxide dispersion strengthened (DSC) and niobium micro-composite, designated and marketed as grade AL15 + Nb1000™. The material was provided by OGM Americas [formerly SCM Metal Products, Research Triangle Park, North Carolina, USA]. The nominal chemical composition of the as-received, or as-provided, material is summarized in Table 1. The copper alloy is strengthened by 0.7 vol% of fine oxide [Al_2O_3]

Table 1 Nominal chemical composition of dispersion strengthened copper-niobium composite [AL15 + NB1000™]

Grade	Al_2O_3 content		Niobium content	
	Weight pct.	Volume pct.	Weight pct.	Volume pct.
AL15 + NB1000™	0.30	0.70	10.0	10.6

particles. The conventional oxide dispersion strengthened copper material was produced by a proprietary powder metallurgy [P/M] internal oxidation process. The internal oxidation process provides a microstructure containing a near uniform distribution of fine particles of aluminium oxide (Al_2O_3). These particles are both chemically stable and thermally stable within the copper matrix. Specific details pertinent to intricacies of the process to obtain the dispersion strengthened copper-niobium micro-composite are considered proprietary by the manufacturer. The two-phase face-centered cubic (f.c.c) body-centered cubic (b.c.c) composite powder was consolidated in the same manner as conventional dispersion strengthened copper (AL-15), into a fully-dense product. The consolidated powder was subsequently finish worked to the desired form using conventional metal working methods. The composite contains 10 pct., niobium (by weight), in the form of uniformly dispersed particles. The resultant micro-composite was subjected to deformation processing by hot extrusion. Specific details pertaining to the intricacies of the technique of ultrasonic nano-crystal Surface Modification can be found elsewhere in the published literature [35] and will not be repeated here.

Experimental Techniques

Initial Microstructure Characterization

Metallographic samples were cut from the as-provided extruded stock of the Cu-Nb micro-composite [AL15 + NB1000™]. The samples were mounted in bakelite and initially wet ground using 320, 400 and 600-grit silicon carbide impregnated emery paper, using copious amounts of water both as lubricant and coolant. Subsequently, the samples were mechanically polished using one-micron alumina-based polishing compound. Grain morphology and other intrinsic microstructural features were revealed using an etchant mixture of 25 ml ammonium hydroxide, 20 ml of hydrogen peroxide [3 pct. concentration] and 25 ml of distilled water. The polished and etched specimens were examined in an optical microscope and photographed using standard bright field imaging technique.

Test Specimen Preparation

Cylindrical test specimens, conforming with specifications detailed in ASTM: E8-2007 [36] were precision machined from the as-provided Cu-Nb micro-composite [AL15 + NB1000™]. The gage section of the test specimens measured 25 mm in length and 6.25 mm in diameter. To minimize the effects and/or influences of surface irregularities and finish, the gage section of the as-provided Cu-Nb micro-composite sample was mechanically ground and then finish polished to remove all the circumferential scratches and surface machining marks.

Indentation Hardness: Microhardness

Hardness is a quantifiable mechanical property of a material that safely provides a measure of the resistance offered by the chosen material, be it metal, ceramic, intermetallic and their composite counterparts, to indentation, densification and cracking [37, 38]. The microhardness test can provide useful information on hardness characteristics of high strength solid metals that cannot be easily determined using a macroscopic hardness test, such as Brinell and Rockwell. In this study, the Vickers microhardness of the samples is measured. The Vickers' microhardness value was determined after precisely measuring size of the indent using a low magnification microscope attached to the indenter [**Model**: INSTRON Wilson Tukon 2100]. The Vickers' indenter creates a square impression from which the length of the two surface projected diagonals was accurately measured. The Vickers' hardness was calculated from the ratio of applied load to the area of contact enclosed within the four faces of the undeformed indenter. The indentation load used was 1000 g (9.8 N) for a dwell time of 15 s. The Hardness number was computed using the expression:

$$\mathbf{H_v} = 1.8545\left[\mathbf{P}/\mathbf{d}^2\right] \tag{1}$$

where **P** is the indentation load used and **d** is the average length of the two diagonals in mm. Two indents were made at different locations along the gage length of each sample and the result is reported as the average value. The micro-hardness measurements (trials and average) made on the surface treated composite sample and untreated composite sample are summarized in Table 2.

Mechanical Testing

Uniaxial tensile tests were performed on test samples of the two chosen composite materials, i.e., as-provided and surface treated, up until failure by deforming the test specimens at a constant strain rate of 10^{-4} s^{-1} at ambient temperature (27 °C). The stress and strain measurements parallel to the load-line were recorded by a PC-based Data Acquisition System (DAS).

Table 2 A compilation of the microhardness measurements made on the untreated and surface treated Copper-Niobium microcomposite

Material	Condition	Trial number	d_1 (μm)	d_2 (μm)	d average (μm)	VHN (Kg/mm^2)	VHN average (Kg/mm^2)
Cu-Nb composite	Surface treated	1	36.16	28.34	32.25	891	984
		2	42.55	16.19	29.37	1076	
	Untreated	1	79.17	47.06	63.11	466	512
		2	77.19	38.18	57.68	557	

Analysis of Failure by Fracture

Fracture surfaces of the fully deformed and failed tensile specimens were carefully examined in a scanning electron microscope (SEM) to:

(i) Determine the macroscopic fracture mode, and
(ii) Characterize the fine scale topography and microscopic mechanisms governing fracture. The distinction between the macroscopic mode and microscopic fracture mechanism is based entirely on the magnification level at which the observations are made. The macroscopic mode refers to the overall nature of failure, while the microscopic mechanisms pertain to failure processes occurring at the fine "microscopic" level, such as: (a) microvoid formation, (b) microscopic void growth and eventual coalescence, and (c) nature of cracking. Samples for observation in the scanning electron microscope were obtained from the deformed and failed tensile specimens by sectioning parallel to the fracture surface.

Results and Discussion

Initial Microstructure

Typical microstructure of the as-provided AL15 + NB1000™ micro-composite, in the extruded condition, for both the longitudinal (L) and transverse (T) orientations are shown in Figs. 1 and 2. The longitudinal (L) microstructure (Fig. 1) reveals the ribbon-like filamentary morphology of the niobium phase following deformation processing. The filamentary morphology and alignment of the reinforcing niobium phase along the extrusion direction is readily apparent. During deformation processing using hot extrusion, the primary and secondary dendrite arms of niobium become aligned and elongated into filaments having a fairly large aspect ratio. The transverse microstructure (Fig. 2) reveals a near-uniform distribution of the reinforcing niobium phase in the oxide dispersion strengthened copper matrix.

Microhardness

Vickers's microhardness measurements were made at independent locations across the gage length of the composite material specimen. Two measurements were made at different locations along the gage length to gather information on hardness. The micro-hardness measurements provide useful information relevant to the intrinsic

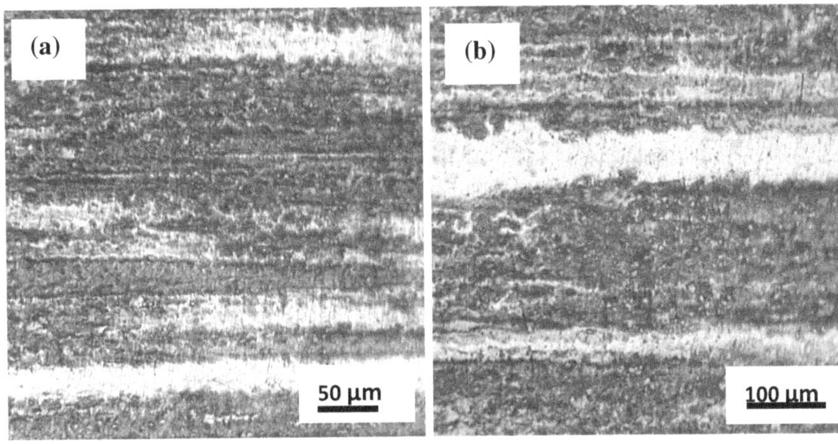

Fig. 1 Optical micrographs illustrating the microstructure of the dispersion strengthened copper-niobium composite in the longitudinal direction

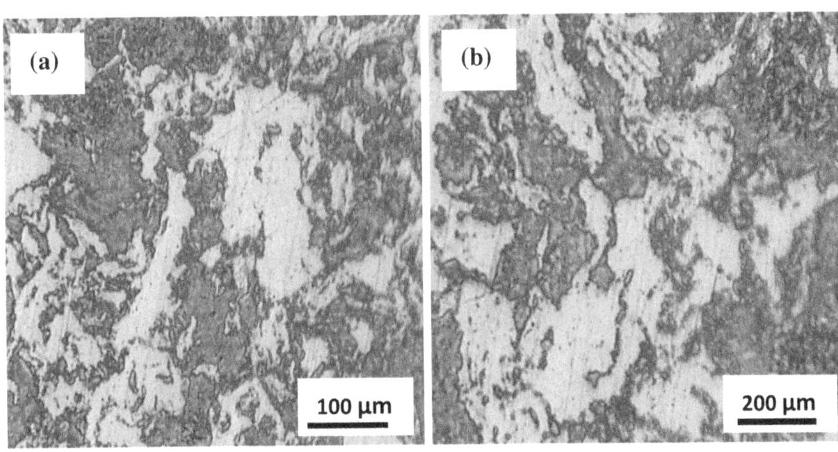

Fig. 2 Optical micrographs illustrating the microstructure of the dispersion strengthened copper-niobium composite in the transverse orientation

influence of size of the grains on strengthening. It is difficult to assess the exact magnitude of strengthening arising from the size and morphology of grains independent of any weakening effect arising from the presence and distribution of defects. The micro-hardness measurements [(i) trials, and (ii) resultant average value (in kg/mm^2)] are summarized in Table 2. Results reveal the micro-hardness of the nanocrystalline surface treated composite to be 90% higher than the as-provided untreated composite counterpart.

Tensile Response

The ambient temperature tensile properties of the AL15 + NB1000™ micro-composite are summarized in Table 3 for both the as-received (untreated) and surface treated counterpart. Results reveal the following key findings:

(i) A noticeable influence of surface treatment on elastic modulus [E] of the as-provided micro-composite [AL-15 + Nb1000™]. The elastic modulus of the untreated sample was 100 GPa, while elastic modulus of the surface treated counterpart was 110 GPa, i.e., 10% higher.
(ii) The ambient temperature yield strength, defined as the stress required at a plastic strain of 0.2 pct., of the Cu-Nb micro-composite was 465 MPa for the as-provided composite and 475 MPa for the surface treated counterpart. A nominal increase of 10 MPa because of the surface treatment.
(iii) The ultimate tensile strength of the as-provided untreated composite is 595 MPa. However, for the surface treated counterpart the ultimate tensile strength is 610 MPa, a nominal increase of 15 MPa or 3.0 pct. The enhanced strength (i.e., yield strength and tensile strength) of the surface treated micro-composite microstructure can be safely attributed to the presence of fine grains on the surface. The locally high dislocation density that is intrinsic to fine grains contributes in a noticeable way to the increase in strength that is observed from a short-time mechanical test, i.e., tension test.
(iv) The engineering strain, or uniform elongation, was 13.5 pct. for the as-provided composite with a marginal drop to 12.5 pct. for the fine grain surface treated counterpart. It is to be noted that the observed increase in strength due to surface treatment occurs without appreciable loss in ductility.
(v) The reduction in test specimen cross-section area, another useful measure to quantify ductility, was 38 pct., for the as-provided untreated composite material and 29 pct., for the surface treated counterpart having fine grains on the surface, a noticeable 20 pct. decrease.

In this study, the dispersion strengthened copper-niobium micro-composite subsequent to surface treatment has marginally higher strength while still maintaining ductility within acceptable limits. The engineering stress versus engineering strain curves for both the untreated and surface treated micro-composite are compared in

Table 3 A compilation of the room temperature tensile properties of the untreated and surface treated Copper-Niobium microcomposite

Material	Condition	Elastic modulus (GPa)	Yield strength (MPa)	Ultimate tensile strength (MPa)	Elongation (%)	Reduction in area (%)
Cu-Nb composite	Surface treated	110	475	610	12.5	29
	Untreated	100	465	595	13.5	38

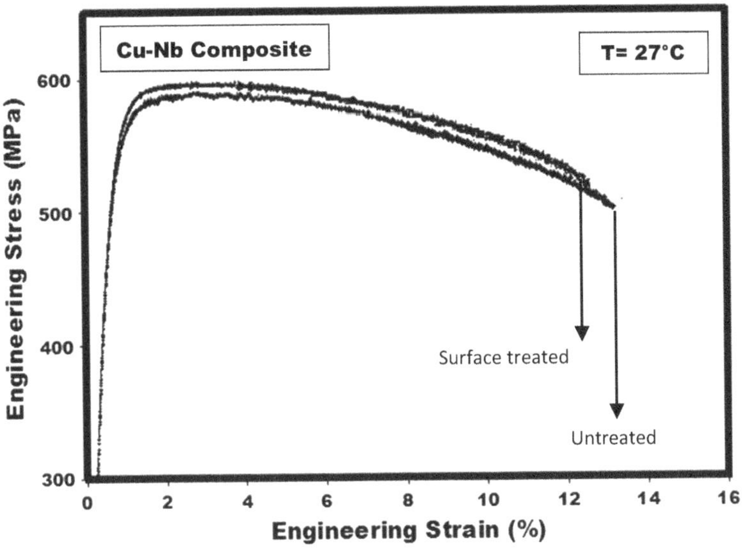

Fig. 3 Influence of surface treatment on the engineering stress versus engineering Strain curve of the copper-niobium composite

Fig. 3. The increased strength and minimal decrease in engineering strain is evident from this figure. A comparison of the monotonic stress versus -strain curves of both the untreated composite and surface treated counterpart is shown in Fig. 4. This

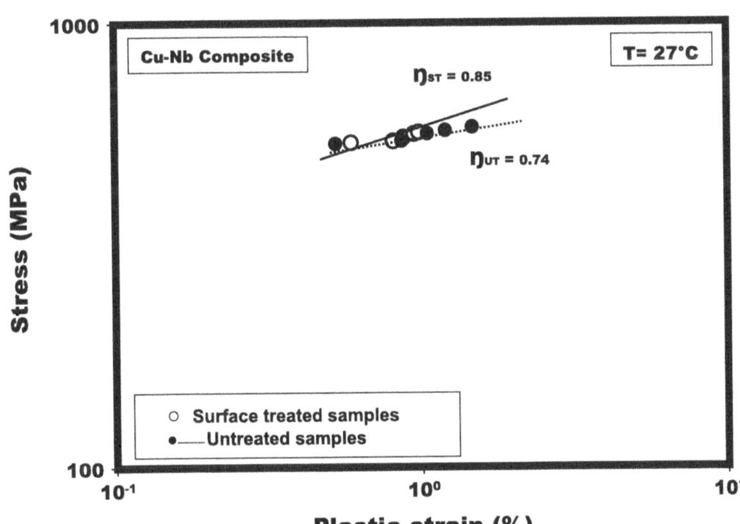

Fig. 4 Comparison of the monotonic stress versus strain curves of the untreated and surface treated composite

figure reveals the degree of strain hardening experienced by the micro-composite whose surface was subject to nanocrystalline treatment to be marginally higher than the untreated counterpart.

Tensile Fracture

Fracture surfaces of the test specimens of both the as-provided, untreated, and surface treated counterpart deformed in uniaxial tension were examined in a scanning electron microscope at:

(a) Low magnification to identify the macroscopic morphology, and
(b) Higher magnifications of the scanning electron microscope to identify the fine-scale fracture features.

Representative fractographs are shown in Figs. 5, 6, 7 and 8.

As-Provided Cu-Nb Micro-composite [AL15 + NB1000^{TM}]

Tensile fracture surface of the as-provided untreated composite sample revealed macroscopic failure to be at an angle to the far-field load axis (Fig. 5a) and overall brittle in appearance. High magnification observation of the fracture surface revealed microscopically rough surface (Fig. 5b). At higher allowable magnifications of the scanning electron microscope the transgranular region was found to be flat and near featureless (Fig. 5c). The region of overload revealed a sizeable population of dimples of varying size intermingled with fine microscopic voids and isolated microscopic cracks, features reminiscent of "locally" ductile and brittle failure mechanisms (Fig. 5d). Macroscopic cracks were distinctly evident in the region separating transgranular failure and overload (Fig. 6a). In the region of overload, at selected locations the fine microscopic voids were found to coalesce to form microscopic cracks during far-field loading (Fig. 6b). The fine microscopic cracks gradually grow through the composite microstructure to become macroscopic cracks.

Surface Treated AL15 + NB1000^{TM} Micro-composite

Tensile fracture surface of the surface treated micro-composite reveals overall failure to be at 45° angle to the far-field stress axis (Fig. 7a). High magnification observation of the fracture surface at the region close to the edge or surface revealed it to be both flat and smooth and inlaid with an array of fine microscopic cracks

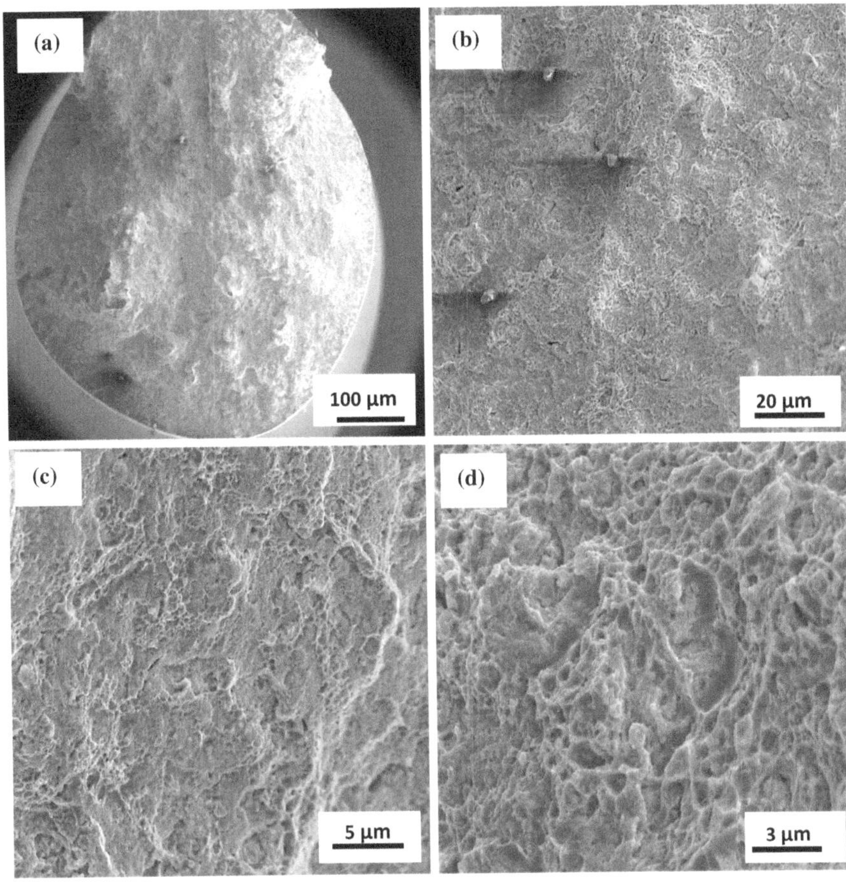

Fig. 5 Scanning electron micrographs of the as-received untreated copper-niobium composite deformed in tension, showing: **a** Overall morphology of failure, **b** High magnification observation of the fracture surface revealing microscopically rough surface, **c** The transgranular region relatively flat and featureless, **d** The region of overload revealing sizeable population of dimples of varying size intermingled with microscopic voids and isolated microscopic cracks

(Fig. 7b). The morphology, orientation and random dispersion of the fine microscopic cracks is shown in Fig. 7c. The region prior to the onset of overload revealed a microscopically rough fracture surface (Fig. 7d). The overload fracture surface revealed few macroscopic cracks intermingled with a random dispersion of fine microscopic cracks and dimples; features indicative of the occurrence of both brittle and ductile failure mechanisms occurring at the "local" level (Fig. 8a). At selected pockets through the overload surface the dimples were well defined along with microscopic void coalescence (Fig. 8b).

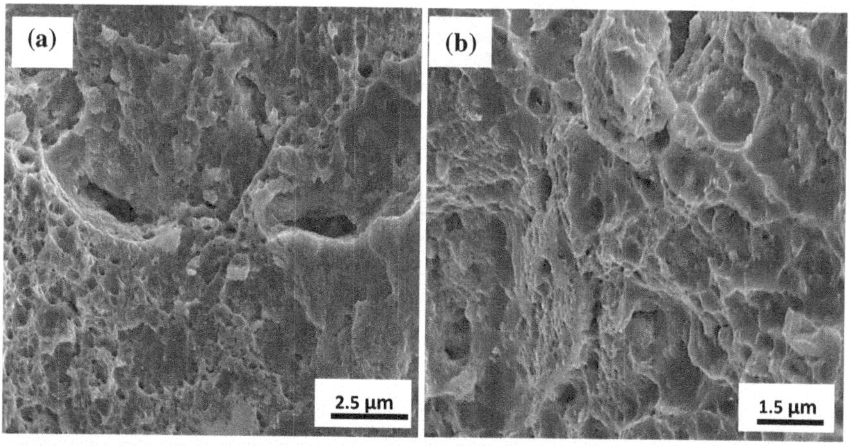

Fig. 6 Scanning electron micrographs of the tensile fracture surface of the as-received untreated copper-niobium composite showing: **a** Macroscopic cracks in the region separating transgranular failure and overload, **b** Fine microscopic void coalescence in the region of overload to form microscopic cracks

Microstructural Influences on Strength-Ductility Relationship

(A) **Influence of Microstructure on Strength**

The observed improvement in strength of the chosen material (i.e. oxide dispersion strengthened copper-niobium micro-composite) because of surface treatment can be attributed to the creation of a gradient in plastic deformation with distance from the treated surface, and thus the generation of a gradient microstructure. At the surface, the occurrence of plastic deformation results in refining the existing grain size to a finer level, approaching the nano-scale. However, well into the interior the influence of plastic deformation, if any, is little and the grain size of the parent composite material is largely unaffected. Thus, strength of the chosen composite can be predicted by adding contributions from reduction in grain size coupled with "local" increase in dislocation density due to plastic deformation [35, 37, 38].

$$\boldsymbol{\sigma_f} = \boldsymbol{\sigma_o} + \mathbf{K}\left[\mathbf{d_p}\right]^{-1/2} + \boldsymbol{\alpha}\mathbf{G}\mathbf{b}\,\boldsymbol{\rho}^{1/2} \tag{2}$$

In this expression σ_f is the strength, σ_O is lattice friction stress that opposes the motion of dislocations, K is the Hall-Petch constant, α is a constant, G is the shear modulus and ρ is density of dislocations and b is the Burger's vector. From this equation, as mean free path of the dislocation decreases with a concurrent increase in the density of dislocations, the hardness of the material increases.

The presence of fine grains both at and immediately beneath the surface does have a beneficial influence on both hardness and strength of metallic materials. The

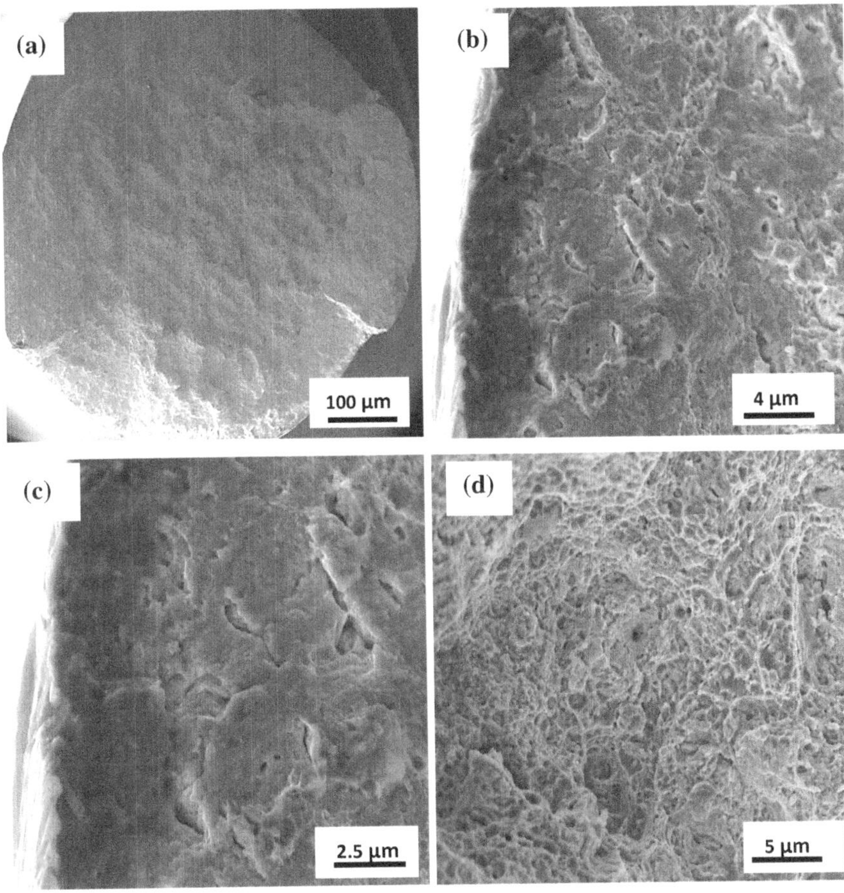

Fig. 7 Scanning electron micrographs of the tensile fracture surface of the surface treated copper-niobium composite deformed in tension, showing: **a** Overall morphology of failure, **b** High magnification observation of the fracture surface at the region of the surface revealing to be flat and microscopically smooth inlaid with an array of fine microscopic cracks. **c** High magnification observation of (**b**) revealing morphology, orientation and dispersion of fine microscopic cracks. **d** The region immediately prior to overload reveals microscopically Rough fracture surface

surface treatment results in the presence of very fine grains at the surface. Thus, due to the existence of a high density of grain boundaries, the mean free path for the movement of dislocations is significantly reduced with a concomitant improvement in strength of the alloy, which was observable for both the yield strength and ultimate tensile strength of the chosen composite material. This is easily quantified by the observed increase in strain hardening capability of the surface treated composite microstructure. The strain hardening exponent for the surface treated composite (η_{ST}) was 0.85 while the strain hardening exponent for the untreated counterpart (η_{UT}) was 0.74.

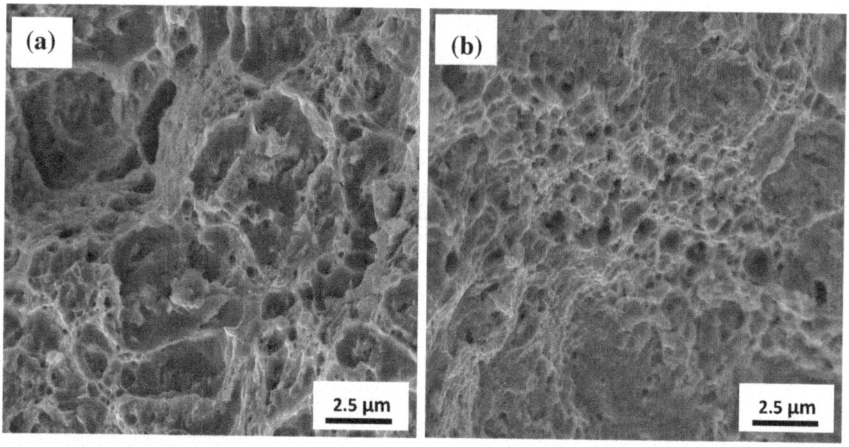

Fig. 8 Scanning electron micrographs of the tensile fracture surface of the surface treated copper-niobium composite showing: **a** Isolate macroscopic cracks, random dispersion of microscopic voids, and dimples on the overload fracture surface, **b** A region on the overload fracture surface revealing observable dimples and microscopic void coalescence

(B) **Influence of Microstructure on Ductility**

It is highly desirable that the chosen material offer a combination of good strength and ductility. Ductility of the chosen copper-niobium micro-composite is governed by its inhibition to necking and the resultant localization of strain. Both are influenced by work hardening capacity of the material. The occurrence of work hardening does prevent localized deformation from occurring thereby delaying the occurrence of necking during a tensile test [39]. Very much in conformance with the criterion of Consider [40, 41] the ductility of a material is essentially determined by its capacity to strain harden, which is the capacity to accumulate dislocations generated during plastic deformation. During plastic deformation that occurs while loading in tension beyond the yield point, the multiplication of dislocations along with their annihilation occurs simultaneously. For the presence of very fine grains at the surface most dislocation sources are not active in operation [42]. Thus, they cannot provide enough dislocations to sustain the plastic deformation. Furthermore, the dislocations cannot be stored primarily because they tend to disappear at the grain boundaries of the fine grains, which often serve as sinks for the annihilation of dislocations [43–45]. This in turn is responsible for the reduced ductility of the fine grain material. The presence of very fine grains at the surface tends to harden the material and this contributes in a small way to improving strength of the material while the relatively unaffected inner layers provide a ductile interior for the accumulation and easy movement of dislocations during tensile deformation.

Microscopic Mechanisms Governing Tensile Fracture

For both the untreated and surface treated dispersion strengthened Cu-Nb micro-composite a combination of:

(a) Fine grain size as a direct result of powder metallurgy processing in conjunction with the presence of fine nano-sized crystals for the surface treated composite, and
(b) The occurrence of an interaction of the mobile dislocations with the Al_2O_3 oxide particles and niobium reinforcement (filaments) serves to homogenize the deformation by facilitating or enabling a near uniform dispersion of dislocations and thereby minimize the occurrence of slip concentration and concomitant localization of strain.

This aids in suppressing the occurrence of inter-crystalline or intergranular deformation that is typical of coarse grained polycrystalline materials.

High magnification observation of the fracture surface essentially revealed a bimodal failure comprising of:

(i) An observable population of voids of varying size, i.e., macroscopic and fine microscopic, and intermingled with pockets of dimples, features reminiscent of the occurrence of locally ductile failure mechanisms.
(ii) An observable number of fine and randomly dispersed microscopic cracks and isolated macroscopic cracks; features reminiscent of locally brittle failure mechanisms.

The microscopic voids formed as a consequence of failure of the reinforcing fine oxide particles by cracking. The voids gradually grow during far-field loading and eventually coalesce. The coalescence of the fine microscopic voids is aided by the formation and presence of void sheets coupled with the intense localization of strain between the growing microscopic voids. The pockets of transgranular region revealed a near-featureless and flat morphology with isolated pockets of shallow dimples indicative of the occurrence of highly localized plastic deformation. The observed distribution of the transgranular regions and 'localized' plastic deformation is dependent on the mutually interactive influences of the following [46, 47]: (i) orientation of grains, (ii) presence and role of grain boundaries, and (iii) grain boundary triple junctions. The observed presence of a population of voids, predominantly microscopic and interdispersed with macroscopic does influence the macroscopic mechanical response of the polycrystalline micro-composite.

Conclusions

An experimental study of the influence of nanocrystalline surface modification of a dispersion strengthened copper-niobium micro-composite provides the following observations:

(1) The longitudinal microstructure revealed ribbon like filaments of the reinforcing niobium phases. The transverse microstructure revealed a non-uniform dispersion of the niobium filaments in the oxide dispersion strengthened copper matrix.
(2) Results reveal the micro-hardness of the nanocrystalline surface treated composite to be 90% higher than the as-provided untreated composite counterpart.
(3) Surface treatment of the as-provided oxide dispersion strengthened copper–niobium micro-composite resulting in a fine grain size revealed a marginal, i.e., 10 pct., increase in elastic modulus and a small yet observable increase in both yield strength and tensile strength. Commensurate with increase in both stiffness and strength, the decrease in ductility quantified by elongation, over a gage length of 12.5 mm, and reduction in test specimen cross-sectional area was minimal.
(4) The surface treated composite material has a multi-later microstructure with fine grains at the surface and a relatively unaffected and ductile interior. The surface layer containing fine grains provides a strong resistance to plastic deformation resulting in marginally higher strength for the surface treated composite. The unaffected interior provides strain hardening capacity and this contributes to maintaining a high ductility.
(5) Tensile fracture of both the as-provided oxide dispersion strengthened Cu-Nb micro-composite and the surface treated counterpart revealed macroscopically ductile failure. At the fine microscopic level failure was predominantly transgranular with isolated regions or pockets containing a random dispersion of fine microscopic cracks coupled with an observable population of microscopic voids of varying size and intermingled with shallow dimples; features reminiscent of locally brittle and ductile failure mechanisms.

Acknowledgements The material used in this research study was provided by OGM Americas (Research Triangle Park, North Carolina, USA). The authors extend gracious and generous thanks to Dr. C. Ye and his students (the University of Akron, Ohio, USA) for doing the surface modification (ultrasonic nanocrystal) on selected samples in their novel research laboratory.

References

1. Nadkarni AV, Klar E, Shafer WM (1976) A new dispersion strengthened copper. Met Eng Q 16(3):10–15
2. Nadkarni AV in E. Liong, Taubenblat PW (eds) High Conductivity Copper and Aluminum Alloys, TMS-AIME, Warrendale, PA, 1984, p 77
3. Morris MA, Morris DG (1989) Microstructural refinement and associated strength of copper alloys obtained by mechanical alloying. Mater Sci Eng 111(1989):115
4. Nadkarni AV, Troxell JD, Verniers F (1989) GlidCop dispersion strengthened copper: an advanced alloy system for automotive and aerospace applications. Report SCM Metal Products, Cleveland, Ohio
5. Troxell JD (1995) Dispersion strengthened Copper-Niobium composites. Adv Mater Process, pp 35–37

6. Zhu F, Jiao L, Wanderka N, Wahi RP, Wallenberger H (1992) FIM atom probe study of an Al_2O_3 dispersion strengthened copper alloy. Surf Sci 266:337–341
7. Jha SC, Delagi RG, Forster JA, Krotz PO (1993) High Strength, high conductivity Cu-Nb microcomposite sheet fabricated via multiple roll bonding. Metall Trans A 24:15–22
8. Hosford WF Jr (1964) Microstructural changes during deformation of [Oil] fiber-textured metals. Trans AIME 230:121–126
9. Verhoven JD, Downing HL, Chumbley LS, Gibson ED (1989) The resistivity and microstructure of heavily drawn Cu-Nb alloys. Appl Phys 65:1293–1299
10. Bevk J, Sunder WA, Dublon G, Cohen DE (1982) In-situ composites IV. In Lemkey FD, Kline HE, McClean M (ed) Elsevier Applied Science Publishers, Amsterdam, p 121
11. Funkenbusch PD, Courtney TH, Kubisch DG (1984) Fabricability of and microstructural development in cold worked metal matrix composites. Scr. Metall 18(10):1099
12. Spitzig WA, Pelton AR, Laabs FC (1987) Characterization of the strength and microstructure of heavily cold worked Cu-Nb composites. Acta Metall 35:2427–2432
13. Funkenbusch PD, Lee JK, Courtney TH (1987) Ductile two-phase alloys: prediction of strengthening at high strains. Metall Trans 18A:1249–1256
14. Bevk J, Harbison JP, Bell JL (1978) Anomalous increase in strength of insitu formed Cu-Nb multifilamentary composites. J Appl Phys 49:6031–6038
15. Spitzig WA, Reed LK, Chatterjee A (1990) Comparison of the low-cycle fatigue properties of heavily cold-drawn copper and Cu-20% Nb. Mater Sci Eng A 123:69–74
16. Zhang HW, Hei ZK, Liu G, Ju J, Lu K (2003) Formation of nanostructured surface layer on AISI 304 stainless steel by means of surface mechanical attrition treatment. Acta Mater 51(7):1871
17. Lu K, Ju J (2004) Nanostructured surface layer on metallic materials induced by surface mechanical attrition treatment. Mater Sci Eng 38A:375–377
18. Chan HL, Ruan HH, Chen AY, Lu J (2010) Optimization of the strain rate to achieve exceptional mechanical properties of 304 stainless steel using high speed ultrasonic surface mechanical attrition treatment. Acta Mater 58(15):5086
19. Liu G, Wang S, Lou X, Lu J, Lu K (2001) Low carbon steel with nanostructured surface layer induced by high-energy shot peening. Scr Mater 44:1791–1799
20. Tao NR, Sui ML, Lu K, Lua K (1999) Surface nanocrystallization of iron induced by ultrasonic shot peening. Nanostruct Mater 11:433–440
21. Liu G, Lu J, Lu K (2000) Surface nanocrystallization of 316L stainless steel induced by ultrasonic shot peening. Mater Sci Eng A 286:91–94
22. Lu JZ, Luo KY, Zhang YK, Lui CY, Sun GF, Zhou JZ, Zhang L, You J, Chen KM, Zhong JW (2010) Grain refinement of LY2 aluminum alloy induced by ultra-high plastic strain during multiple laser shock processing impacts. Acta Mater 58:3984–3994
23. Lu JZ, Luo KY, Zhang YK, Sun GF, Gu YY, Zhou JZ, Ren XD, Zhang XC, Zhang LF, Chen KM, Cui CY, Jiang YF, Feng AX, Zhang L (2010) Grain refinement mechanism of multiple laser shock processing impacts on ANSI 304 stainless steel. Acta Mater 58:5354–5360
24. Ye C, Liao Y, Cheng GJ (2010) Warm laser shock peening driven nanostructures and their effects on fatigue performance in aluminum alloy 6160. Adv Eng Mater 14:291–295
25. Ye C, Suslov S, Kim BJ, Stach EA, Cheng GJ (2011) Fatigue performance improvement in AISI 4140 steel by dynamic strain aging and dynamic precipitation during warm laser shock peening. Acta Mater 59:1014–1020
26. Gill AS, Zhou Z, Lienert U, Almer J, Lahrman DF, Mannava SR, Qian D, Vasudevan VK (2012) High spatial resolution, high energy synchrotron x-ray diffraction characterization of residual strains and stresses in laser shock peened Inconel 718 alloy. J Appl Phys, vol 111
27. Lu JZ, Qi H, Luo KY, Luo M, Cheng XN (2014) Corrosion behaviour of AISI 304 stainless steel subjected to massive laser shock peening impact with different pulse energies. Corros Sci 80:53–60
28. Luo KY, Lu JZ, Wang QW, Luo M, Qi H, Zhou JZ (2013) Residual stress distribution of Ti-6Al-4V alloy under different ns-LSP processing parameters. Appl Surf Sci 285:607

29. Wang ZB, Lu J, Lu K (2006) Wear and corrosion properties of a low carbon steel processed by means of SMAT followed by lower temperature chromizing treatment. Surf Coat Technol 201:2796–2799
30. Lu JZ, Luo KY, Dai FZ, Zhong JW, Xu LZ, Yang CJ, Zhang L, Wang QW, Zhong JS, Yang DK, Zhang YK (2012) Mater Sci Eng 536 A:57–67
31. Jelliti S, Richard C, Retraint D, Roland T, Chemkhi M, Demangel C (2013) Effect of surface nanocrystallization on the corrosion behavior of Ti–6Al–4V titanium alloy. Surf Coat Technol 224:82–92
32. Faghihi S, Zhilyaev AP, Szpunar JA, Azari F, Vali H, Tabrizian M (2007) Nanostructuring of a titanium material by high-pressure torsion improves pre-osteoblast attachment. Adv Mater 19(8):1069–1073
33. Pyun YS, Cho IH, Suh CM, Park J, Rogers J, Kayumov R, Murakami R (2013) Reducing production loss by prolonging service life of rolling mill shear pin with ultrasonic nanocrystal surface modification technology. Int J Precis Eng Man 14(11):2027–2032
34. Cherif C, Pyoun Y, Scholtes B (2009) Effects of ultrasonic nanocrystal surface modification (UNSM) on residual stress state and fatigue strength of AISI 304. J Mater Eng Perform 19 (2):282–286
35. Ye C, Telang A, Gill AS, Suslov S, Idell Y, Zweiacker K, Wiezorek JMK, Zhou Z, Qian D, Mannava SR, Vasudevan VK (2014) Gradient nanostructure and residual stresses induced by ultrasonic nano-crystal surface modification in 304 austenitic stainless steel for high strength and high ductility. Mater Sci Eng 613 A:274–288
36. American Society for Testing Materials (1998) Standard E-8-98, Standard method for tension testing of metals. ASTM, Philadelphia, PA, USA
37. Bruet BJF, Song J, Boyce MC, Ortiz C (2008) Materials design principles of ancient fish armour. Nat Mater 7:748–753
38. Chen AY, Ryan HH, Wang J, Chan HL, Wang Q, Li Q, Lu J (2011) Acta Mater 59: 3697–3702
39. Fang TH, Li WL, Tao NR, Lu K (2011) Revealing extraordinary intrinsic tensile plasticity in gradient nano-grained copper. Science 331:1587
40. Dieter G (1986) Mechanical metallurgy, 3rd edn. McGraw Hill Science, New York
41. Wang YM, Ma E (2004) Strain hardening, strain rate sensitivity, and ductility of nanostructured metals. Materials Science and Engineering 46:375–377
42. Lu K, Lu L, Suresh S (2009) Strengthening materials by engineering coherent internal boundaries at the Nanoscale. Science 324:349
43. Wang YM, Ma E (2004) Three strategies to achieve uniform tensile deformation in a nanostructured metal. Acta Mater 52:1699–1710
44. Ma E (2003) Instabilities and ductility of nanocrystalline and ultrafine-grained metals. Scr Mater 49(7):663
45. Meyers MA, Mishra A, Bendon DJ (2006) Mechanical properties of nanocrystalline materials. Prog Mater Sci 51(4):427
46. Srivatsan TS, Naruka AS, Ravi BG, Sudarshan TS, Petraroli M, Riester L (2002) Microstructure and properties of molybdenumprinciple-copper composite metal samples consolidated by plasma pressure compaction. Powder Metall 45(3):255–260
47. Srivatsan TS, Ravi BG, Naruka AS, Petraroli M, Kalyanaraman M, Sudarshan TS (2002) nfluence of consolidation parameters on the microstructure and hardness of bulk copper samples made from nanopowders. Mater Des 23:291–296

Fundamental Issues and Highlights of Reactive Wetting in Carbon-Based Composites

Khurram Iqbal and Stevens Cadet

Abstract Wetting can be broadly classified into two categories: (a) non-reactive wetting (b) and reactive wetting. A chemical reaction occurs between the liquid/solid interface, and the resulting chemical bonds are responsible for wetting. The recent results and theoretical developments concerning the reactive wetting of carbon by liquid copper alloys are studied. A section is devoted to illustrate and discuss the effect of interfacial reactions in carbon/copper alloys system.

Keywords C/Cu composites · Reactivity · Wetting

Introduction

This paper contains a comparative study of fundamental issues of reactive wetting as well as reported fabrication techniques of carbon based liquid copper alloys (Fig. 1). These composites have become very popular among researchers for high temperature applications because of the high thermal conductivity offered by the copper matrix [1–10]. These types of advanced composite materials are generally manufactured by liquid state routes consisting of infiltration of the liquid-copper alloys into solid-carbon preforms. Reactive melt infiltration has emerged as a viable approach for fabrication of advanced composite materials with quick processing time, low cost and net or near-net shape components that is difficult to machine. The reinforcement- carbon preforms in situ via a chemical reaction during infiltration,

K. Iqbal (✉)
Department of Mathematical Sciences, Federal Urdu University of Arts, Sciences and Technology, Karachi 75300, Pakistan
e-mail: khurramiqbal.nust@gmail.com

K. Iqbal · S. Cadet
Architect Industries Lab, 170 Hathaway Avenue Elmont, New York 11003, USA

T. S. Srivatsan et al. (eds.), *Metal-Matrix Composites Innovations, Advances and Applications*, The Minerals, Metals & Materials Series,
https://doi.org/10.1007/978-3-319-72853-7_15

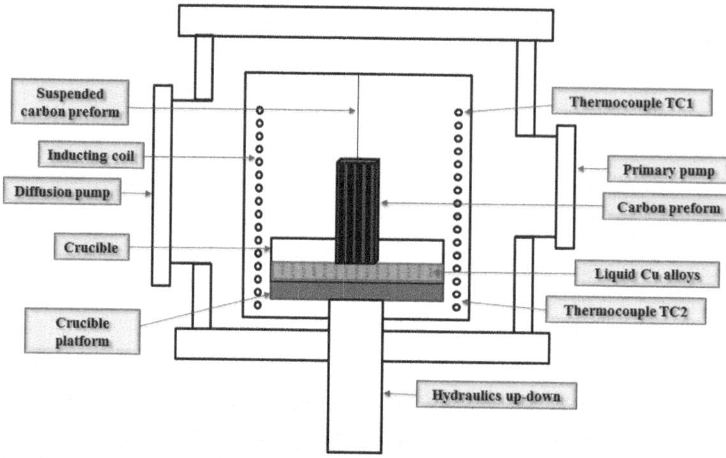

Fig. 1 Schematic view of the experimental setup

which can be controlled to achieve the desired level of conversion and structure [11–15]. The basic physics and modeling of reactive flows through porous media, in the context of reactive infiltration, are discussed [15–27]. There are two types of infiltration. If the liquid does not wet the solid (Fig. 2) infiltration must be performed using pressure to overcome the capillary pressure. In contrast, in wetting systems, spontaneous infiltration can occur. Molten copper alloys wet carbon substrates and for this reason they can spontaneously infiltrate carbon preforms. As a result of the high thermal stability and low specific mass of these composites, these systems have great importance for space radiator materials [11, 28–35]. As the wettability between copper and carbon is poor [10, 13, 36–39], the addition of silicon or titanium or chromium to copper as an alternative element to lead has been discussed to improve wettability [13, 15–27, 40].

From the recent published research papers on the subject of reactive wetting [33–41] it appears complexity, and no fundamental approach has been developed on spreading kinetics in reactive systems. This paper presents the main results obtained of the dynamic wetting effect.

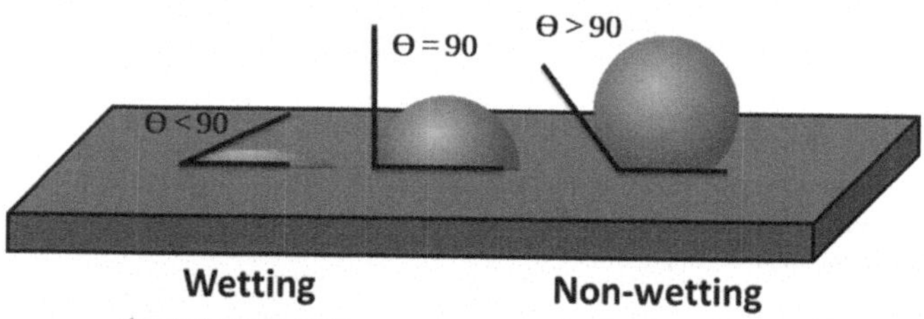

Fig. 2 The wetting phenomenon of liquids on a carbon substrate

Thermodynamics of Silicon, Titanium and Chromium Carbides Formations

The stoichiometrical conversion of carbon to silicon carbide (SiC) [12, 31–34] is shown in Eq. 1.

$$\mathrm{Si}(l) + C(s) \rightarrow SiC(s) \quad \Delta H = -68\,\mathrm{kJ/mol} \tag{1}$$

That means, the formation of SiC with a reaction enthalpy of −68 kJ/mol is highly exothermic.

For the liquid copper-titanium alloys react very fast with carbon to form solid titanium carbides (TiC) on its surface [12–15, 23], and the possible reaction and accompanying free-energy for the formation of TiC are shown in Eqs. 2 and 3, respectively.

$$yCuTi(liquid) + C \rightarrow TiC_{1-x} + Cu_yTi_{y-1} + xC \tag{2}$$

$$\Delta G^0 = -186606 + 13.22T\left[\mathrm{Jmol^{-1}}\right] \tag{3}$$

where ΔG^0 is the standard Gibbs energy of the reaction and T is the temperature expressed in Kelvin.

Based on Lu et al. [14, 15, 23], the changes in enthalpy (ΔH^0) and Gibbs free energy of formation of TiC can be divided into two temperature ranges:

when T < 1939 K,

$$\Delta H^0 = -184571.8 + 5.042T - 2.425 \times 10^{-3}T^2 - 1.958 \times 10^6/T \tag{4}$$

$$\Delta G^0 = -184571.8 + 41.382T - 5.042T\,\ln T + 2.425 \times 10^{-3}T^2 - 9.79 \times 10^5/T \tag{5}$$

and when T $\geq$ 1939 K,

$$\Delta H^0 = -160311.5 + 24.79T - 2.732 \times 10^{-3}T^2 - 1.862 \times 10^6/T \tag{6}$$

$$\Delta G^0 = -160311.5 - 186.97T - 24.79T\,\ln T + 2.732 \times 10^{-3}T^2 - 9.31 \times 10^5/T \tag{7}$$

It can be seen that the Gibbs energy is always negative, indicating the possibility of formation of TiC. Titanium activity a_{Ti} depends only on temperature and can be expressed from the values of the Gibbs energy of formation of TiC as shown in Eq. 8.

Table 1 Free energies of formation of Cr-carbide phases at 1130 °C

Reaction	ΔG/(cal/g.atom)
$\frac{3}{5}Cr + \frac{2}{5}C = \frac{1}{5}Cr_3C_2$	−4594
$\frac{7}{10}Cr + \frac{3}{10}C = \frac{1}{10}Cr_7C_3$	−4065

$$a_{Ti} = \exp\left(\frac{\Delta G^0(T)}{RT}\right) \tag{8}$$

For chromium-carbide phases at 1130 °C, free energies of formation, based upon a gram-atom [11, 15, 25–27], are listed in Table 1. The reaction layer phase that forms at Gr/Cu-Cr alloy interface is Cr_3C_2 and, this result is consistent with thermodynamic considerations of the most stable reaction layer phase. Assuming Henrian behavior, for the Cu-1.22 wt Cr alloy [25–29], the activity of chromium in liquid copper is found to be 0.61.

Reactive Wetting

During the last decades, significant improvements have been made in the measurement of contact angles of high temperature systems [5–14, 32–35, 41]. Wetting studies have benefited from high resolution techniques for characterizing the topological and chemical features of surfaces at nanometric scale. Another reason for this improvement has been the use of monocrystalline or vitreous solids to prepare the high-quality surfaces required for Young's contact angle determinations. Over the past 15 years, further improvements have been made as an increasing number of laboratories are now using more sophisticated versions of the sessile drop method. Sessile drop method is a powerful tool for the study of the transient stages of initial solid-liquid contact and its subsequent evolution, which is a result of the minimization of interfacial energy. Finally, automatic systems for data acquisition and image analysis leading to the simultaneous measurement of contact angle and surface tension have been developed and used widely.

Data on the wetting of carbon materials by liquid metals published until the end of the last century have been extensively reviewed in [5–20, 40, 41]. After giving a brief overview of the fundamental equations of wetting and adhesion for both smooth and rough solid surfaces, this article aims to review results obtained during the last 15 years for non-reactive and reactive liquid–solid systems. Although this review focuses on liquid metal/ceramic systems, a limited number of results concerning liquid metal–solid metal systems are also given because they are useful in understanding bonding at metal/ceramic interfaces. When a pure liquid wets an inert solid surface, the wetting driving force at time *t* [10, 18–22, 26, 31] is given by:

$$F_d(t) = \sigma^0_{SV} - \sigma^0_{SL} - \sigma^0_{LV}\cos\theta(t) \tag{9}$$

where σ^0_{SV}, σ^0_{SL} and $\Delta G(t)$ are the surface energies of the system, and $\theta(t)$ is the instantaneous contact angle. Aksay et al. [1–3] proposed that the dynamic contact angle changes according to Eq. 10.

$$\cos\theta(t) = \cos\theta^0 - \frac{\Delta\sigma(t)}{\sigma^0_{LV}} - \frac{\Delta G(t)}{\sigma^0_{LV}} \tag{10}$$

where $\theta(t)$ and θ^0 denote the contact angles when $t = t(s)$ and $t = 0(s)$, respectively.

σ^0_{LV} and $\Delta\sigma(t)$ are the surface tension of the liquid alloy and the changes in the interfacial tension, respectively. $\Delta G(t)$ is the change in the Gibbs energy due to the dissolution of the solute elements. They argued that the effect of the last term is the strongest during the early stage of wetting, because the interfacial reaction rate is at its maximum when the liquid contacts with the unreacted solid surface. Thereafter, the reaction kinetics slow down, and after an initial decrease, the contact angle increases and gradually approaches the equilibrium value as shown in Fig. 1. Naidich performed calculations of $\Delta G(t)$, and is estimated by [18–22]:

$$\Delta G(t) = \int_0^{\alpha_{eq}} \Delta G_R d\alpha \tag{11}$$

where α and α_{eq} are the current and equilibrium degrees of progress of the reaction and ΔG_R the Gibbs free energy of the reaction per unit area, respectively.

For infiltration to take place, the porous carbon should have pores that are open and interconnected to ensure that the liquid fully infiltrates into the carbon pores and that a completely dense product is obtained [31–36]. During the infiltration process, the pores present in the preform are filled by a matrix metal alloy [34–38]. Since the wettability between copper and carbon is poor [10–17], neither does liquid copper wet carbon nor does chemical reaction occurs. This low wettability results often leads to a debonding during thermal cycling. A small amount of silicon or chromium or titanium additions into Cu can be effective for increasing the wettability between copper and carbon. Then, the wetting angle decreases from 137° for the pure copper to 36° for the alloy Cu-25 at. % silicon, 40° for the alloy Cu-1 at. % chromium and 130° for the alloy Cu-10 at. % titanium [21–32]. Wetting occurred only for non-stoichiometric titanium carbide with x $\leq$ 0.65 was observed [24–26]. A copper alloy with 1 at. % chromium was found to wet chromium carbide substrates [15, 26]. Also in the early 1970s, interaction between copper atoms and the graphite basal plane to be probably no more than a van der Waals force was showed [15, 24–27]. More recently, ceramic fiber-reinforced metals, and graphite-reinforced copper have been reviewed [1–3], with interfacial problems categorized into "chemical" and "mechanical" types. The copper-graphite system again is identified primarily by its lack of chemical reaction, with mention of the addition to copper of 1 at. % of molybdenum, vanadium, manganese, tungsten, iron, or cobalt as active carbide-forming elements for better wetting characteristics [21–27].

Poor wetting is generally observed in non-reactive metal/carbon system. Indeed, in this system the angle θ formed at the contact line of the three phases, solid, liquid and vapor, is usually higher than 90° [1–3]. Contact angle between copper and carbon under vacuum is as high as 120°–140° [1–3]. A considerable improvement in wetting can be produced using certain alloying elements. The contact angle was assumed to exponentially decrease with time, the wetting of carbon by adding small amounts of active metals into copper were measured (Fig. 3) with the following empirical equation [33–36, 39–41]:

$$\theta(t) = \theta_\infty + \theta_\infty \exp(B - At) \tag{12}$$

Here A and B are material constants, and $\theta(t)$ and θ_∞ are the instantaneous and equilibrium values of contact angle. The constant values (A and B) were calculated (Table 2) using the values of θ_0 and θ_F, and reported in [34–36, 39, 40].

From Table 2, the addition of chromium or massive amounts of titanium promotes wetting. Ti as well as Cr reduces the contact angle and the wetting of graphite appears at 1100 °C with Cu-17.5%wt Ti and Cu-28%wt Ti alloys after approx. 100 and 300 s, or at 1200 °C with Cu-1%wt Cr alloy after 100 s, respectively [11, 12, 15, 23, 31, 34]. The reaction rate in the reactive systems increases when the titanium or chromium mole fraction increases. However, this variation in reaction rate has no effect on the final contact angle. From these two facts, it can be concluded that the contribution of the $\Delta G(t)$ term (Eq. 10) to the driving force of wetting is negligible in comparison to the contribution of the change in interfacial energies produced by the reaction.

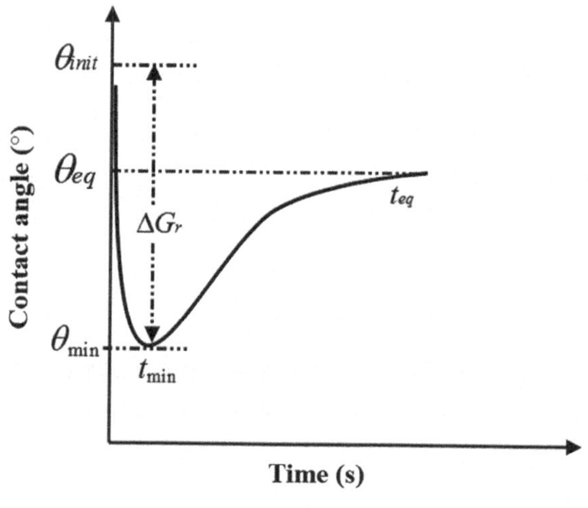

Fig. 3 Contact angle versus time curve in a reactive system

Table 2 Material systems and properties used in this study

	θ_∞ (degrees)	A (/sec)	B
Cu-25%wt Si	36	0.0025	1.4939
Cu-1%wt Cr	20	0.0033	1.8281
Cu-10%wt Cr	14	0.0014	1.7394
Cu-10%wt Ti	118	0.0006	−2.2899
Cu-17.5%wt Ti	6.5	0.0046	3.07413
Cu-28%wt Ti	3	0.0017	2.8247

In summary, the results for the wetting of carbon by copper alloys, therefore, can be interpreted in terms of the physical properties of the reaction product carbides and carbide/metal interfaces. Numerical measurement on the wetting of carbides by Ti and Cr, may be of considerable assistance in the selection of alloys to be used in the production of carbon fiber composites by the infiltration technique. For liquid state processing synthesis of composite materials based on Cu/C system, addition of Si, Ti and Cr into the copper is an effective method to increase the wettability between copper and carbon. Si, Ti and Cr as active alloying elements to copper promote reactive wetting between graphite and copper due to the formation of their carbides at interfaces. The new developments would take into account the effects on wetting kinetics. The liquid is considered to wet, the solid completely only when the contact angle closes to zero.

References

1. Li Y, Bai P, Li Y (2009) Fabrication and fibre matrix interface characteristics of Cu/C(Fe) composite. Sci Sinter 41:193–198
2. Hua Z, Liu Y, Yao G, Wang L, Ma J, Liang L (2010) Preparation and characterization of nickel-coated carbon fibers by electroplating. J Mater Eng Perform 21:324–330
3. Liu Y, Zhang C, Qiao S, Yang Z (2010) Fabrication and microstructure of C/Cu composites. Adv. Eng. Mater 12. https://doi.org/10.1002/adem.200900288
4. Mizumoto M, Tajima Y, Kagawa A (2004) Thermal expansion behavior of SiCP/aluminum alloy composites fabricated by a low-pressure infiltration process. Mater Trans 45:1769–1773
5. Lin MH, Buchgraber W, Korb G, Kao PW (2002) Thermal cycling induced deformation and damage in carbon fiber reinforced copper composite. Scr. Mater. 46:169–173
6. Mayerhofer KE, Neubauer E, Sittner CE, Hutter H (2002) Adhesion promotion of Cu on C by Cr intermediate layers investigated by the SIMS method. Anal Bioanal Chem 374:602–607
7. Eustathopoulos N (1998) Dynamics of wetting in reaction metal/ceramic systems. Acta Mater 46:2319–2327
8. Michaud V, Mortensen A (2001) Infiltration processing of fibre reinforced composites: governing phenomena. Compos Part A: Appl Sci Manuf 32:981–996
9. Neubauer E, Chotikaprakhan S, Dietzel D, Bein BK, Pelzl J, Eisenmenger-Sittner C, Schrank C, Korb G (2006) Loss of adhesion strength of PVD Cu films on carbon substrates after heat treatment and correlated effects on the thermal interface properties. J Appl Surf Sci 252:5432–5436
10. Standing R, Nicholas M (1978) The wetting of alumina and vitreous carbon by Cu–Sn–Ti alloys. J Mater Sci 13:1509–1514

11. Abel PB, Andras LK, Frank SH, Stephen VP (1994) Study of copper on graphite with titanium or chromium bond layer. J Mater Res 9:617–624
12. Dezellus O, Eustathopoulos N (2010) Fundamental issues of reactive wetting by liquid metals. J Mater Sci 45:4256–4264
13. Iqbal K, Sha JJ, Lei ZK, Maqsood A, Mujahid M (2014) Numerical studies of infiltration dynamics of liquid-copper and silicon/solid-carbon system. JOM 66:953–959
14. Shinoda T, Liu H, Mishima Y, Suzuki T (1991) Interfacial compatibility in ceramic-fibre-reinforced metal composites. Mater Sci Eng 146:91–104
15. Dezellus O, Hodaj F, Eustathopoulos N (2003) Progress in modeling of chemical-reaction limited wetting. J Eur Ceram Soc 23:2797–2803
16. Arthur JR, Cho AY (1973) Adsorption and desorption kinetics of Cu and Au on (0001) graphite. Surf Sci 36:641–660
17. Fitzer E (1988) Composites for high temperatures. Pure Appl Chem 60:287–302
18. Kim T, Lee J, Kim Y, Kim JM, Yuan Z (2009) Investigation of the dynamic reactive wetting of Sn–Ag–Cu solder alloys on Ni(P)/ Au coated Cu substrates. Mater Trans 50:2695–2698
19. Zhou ZM, Gao J, Li F, Wang YP, Kolbe M (2011) Experimental determination and thermodynamic modeling of phase equilibria in the Cu–Cr system. J Mater Sci 46:7039–7045
20. Zhan Y, Peng D, She J (2012) Phase equilibria of the Cu–Ti–Er system at 773 K (500 °C) and stability of the $CuTi_3$ phase. Metall Mater Trans A 43A:4015–4022
21. Lü L, Fuh JYH, Wong YS (2001) Characterization modeling and optimization. In: Laser-induced materials and processes for rapid prototyping. Springer, Boston, MA, p 201–239
22. Aksay I, Hoye C, Pask J, Lupis CHP (1974) Chemical thermodynamics of materials. J Phys Chem 78:1178
23. Naidich YV (1981) The wettability of solids by liquid metals. Prog Surf Membr Sci 14: 353–484
24. Zhou XB, De Hosson JM (1996) Reactive wetting of liquid metals on ceramic substrates. Acta Mater 44:421–426
25. Mortimer DA, Nicholas M (1970) The wetting of carbon by copper and copper alloys. J Mater Sci 5:149–155
26. Murr L (1974) Interfacial phenomenon in metals and alloys, Reading MA: Addison-Wesley Publishing Co, Boston, p 102
27. Masoudi M, Hashim M, Kamari HM, Salit MS (2013) Fabrication and characterization of Ni–SiC–Cr nanocomposite coatings. Appl Nanosci 3:357–362
28. Aghaie E, Najafi A, Maleki-Ghaleh H, Mohebi H (2013) Effect of SiC concentration in electrolyte on Ni–SiC composite coating properties. J Surf Eng 29:177–182
29. Bahaaideen FB, Ripin ZM, Ahmad ZA (2010) Electroless Ni-P-C_g (graphite)-SiC composite coating and its application onto piston rings of a small two stroke utility engine. JSIR 69:830–834
30. Vaezi MR, Sadrnezhaad SK, Nikzad L (2008) Electrodeposition of Ni-SiC nano-composite coatings and evaluation of wear and corrosion resistance and electroplating charateristics. Colloids Surf A 315:176–182
31. Bougiouri V, Voytovych R, Dezellus O, Eustathopoulos N (2007) Wetting and reactivity in Ni-Si/C system: experiments versus model predictions. J Mater Sci 42:2016–2023
32. Sommers A, Wang Q, Han X, Joen CT, Park y, Jacobi A (2010) Ceramics and ceramic matrix composites for heat exchangers in advanced thermal systems—a review. Appl Therm Eng 30:1277–1291
33. Yang X, Zhao-hui C, Feng C (2014) High-temperature protective coatings for C/SiC composites. J Asian Ceram Soc 2:305–309
34. Kumar S, Kumar A, Mala RB, Mokhasunavisu RR (2015) Fabrication and ablation studies of 4D C/SiC composite nozzle under liquid propulsion. Int J Appl Ceram Tec 12:176–190
35. Srivastava VK (2012) Micro-structural characterization of Si-SiC ceramic derived from C/C-SiC composite. Am J Mater Sci 2:1–4

36. Kumar S, Kumar A, Devi R, Shukla A, Gupta AK (2009) Capillary infiltration studies of liquids into 3D-stiched C-C performs part B: kinetics of silicon infiltration. J Eur Ceram Soc 29:2651–2657
37. Margiotta JC, Zhang D, Nagle DC (2010) Microstructural evolution during silicon carbide (SiC) formation by liquid silicon infiltration using optical microscopy. Int J Refrac Met Hard Mater 28:191–197
38. Sangsuwan P, Tewari SN, Gatica JE, Singh M, Dickerson R (1999) Reactive infiltration of silicon melt through microporous amorphous carbon preforms. Metall Mater Trans B 30B:933–944
39. Yang J, Ilegbusi OJ (2000) Kinetics of silicon-metal alloy infiltration into porous carbon. Compos Part A 31:617–625
40. Asthana R (2000) Dissolutive capillary penetration with expanding pores and transient contact angles. J Colloid Interf Sci 231:398–400
41. Asthana R (1998) Dynamic wetting effects during infiltration of metals. Scripta Mater 38:1203–1210

Part V
Poster Session

Influence of Graphene Nanoplatelet Reinforcements on Microstructural Development and Wear Behavior of an Aluminum Alloy Nanocomposite

Mohammad Alipour, Reza Eslami Farsani and Yu. A. Abuzin

Abstract Microstructure and wear behavior of aluminum alloy AA7068/graphene nanoplate composites produced by ball milling, stir casting and ultrasonic waves have been investigated. The microstructural studies of the alloy revealed that graphene nanoplatelet addition reduces the grain size, but adding higher graphene nanoplatelet content (1 wt% graphene nanoplatelet) does not change the grain size considerably. T6 heat treatment was applied for all specimens before wear testing. Significant improvements in wear behavior were obtained with the addition of graphene nanoplatelet combined with T6 heat treatment. At higher graphene nanoplatelet contents, the presence of graphene agglomerate on grain boundaries was found to be the favored path for crack growth. The optimum amount of nanoparticles is 0.5 wt% graphene nanoplatelet. Dry sliding wear performance of the alloy was examined in normal atmospheric conditions. The experimental results showed that the T6 heat treatment considerably improved the resistance of 7068 aluminum alloy reinforced with 0.5 wt% graphene nanoplatelet to the dry sliding wear.

Keywords Metal matrix nano composites (MMNCs) · Mechanical properties
Microstructures · Powder processing · Ultrasonic waves

Introduction

The composition of high strength and ductility makes aluminum a special metal for energy absorption, and the high strength aluminum alloys are therefore fine suitable in car systems designed to absorb energy in a clash. In the sketching of wrought

M. Alipour (✉) · R. E. Farsani
Faculty of Materials Science and Engineering, K. N. Toosi University of Technology, Tehran, Iran
e-mail: alipourmo@ut.ac.ir

Yu. A. Abuzin
Faculty of Materials Science and Engineering, National University of Science and Technology (MISIS), Moscow, Russia

T. S. Srivatsan et al. (eds.), *Metal-Matrix Composites Innovations, Advances and Applications*, The Minerals, Metals & Materials Series,
https://doi.org/10.1007/978-3-319-72853-7_16

super high strength aluminum alloys, some of the important factors and properties to be considered are chemical composition and processing parameters and the resulting effects of the micro and macrostructure on the mechanical properties. The cast super high strength aluminum alloys rely on the design of proper chemical composition followed by proper the heat treatment to develop the designed microstructures and properties. Cast aluminum alloys based on the Al–Zn–Mg–Cu system respond very favorably to age hardening and possess a high specific strength [1]. High strength aluminum alloy have been widely used as structural materials in aircraft structure applications due to their attractive comprehensive properties, such as low density, high strength, ductility, toughness and resistance to the fatigue [2]. Many investigations have been carried out on the properties of the 7XXX series aluminum alloy, which show that the mechanical properties is influenced by a range of microstructural features, such as the coarse particles [3], the configuration of the precipitation free zone and the grain size, and aging conditions [4–6].

Super high strength aluminum alloys have been extensively studied after mechanical deformation for several decades [5–7], but little attention has been made on the alloy in as-cast condition and semi-solid state. As-cast structures of these alloys have a significant influence on their mechanical properties and the quality of finished products [6]. The structure of such materials can be controlled by some important factors such as: changing the composition, adding grain refining agents, minimizing inclusions and applying thermomechanical treatments [7]. The use of high concentrations of alloying elements results inhomogeneity in the microstructure and severe segregation of second phases. In casting products, the mechanical properties may vary from location to location due to the variations of grain size, the amount of eutectic phases and the amount of precipitates. Much attention has been made to reduce the segregation of the alloying elements during solidification period of high-alloyed Al alloys [5–9].

Application of metal matrix composites in such industries as automobile, military and aviation has been the subject of many studies in the past two decades [7–9]. The results have shown that when compared with monolithic alloys, metal matrix composites have advantages including better resistance to fatigue, wear and creep as well as higher strength. Metal-matrix nanocomposites based on lightweight metals such as Mg and Al are being investigated in the hopes of producing strengths comparable to those of much compactor metals [9]. The low ductility of metal matrix composites is generally due to the presence of a high percentage of brittle and coarse reinforcement phase in the matrix. It has been shown that properties of nanocomposites can be greatly improved with incorporation of even small amounts of nanoparticles.

Since poor tribological performance limits the use of aluminium and its alloys in wear related applications, many efforts including modification of bulk and surface properties have been made to improve their wear and corrosion wear resistance. Recently Alipour et al. have investigated the effect of aging temperature on the properties of Al–12Zn–3Mg–2.5Cu alloy.

In these investigations ultrasonic powers higher than 1500 W have been employed and uniformity of distribution of graphene plates has been studied. This study investigates the microstructure and wear behavior of AA7068 aluminum alloy/graphene nanocomposites using milling of graphene with aluminum powders and mechanical mixing and casting methods.

Experimental Procedures

The materials used in this investigation were 99% pure aluminum powder with an average particle size of 45 μm and graphene nanoplatelets with an average thickness of approximately 5–10 layer and an average diameter of 3–25 μm. Figure 1 shows the SEM micrographs of Al powder and graphene nanoplatelets. To produce nanocomposites, the graphene nanoplatelets were dispersed in 99.5% ethanol by ultrasonication. The aluminum powder and the reinforcement slurry were added to ball milling for 2 h at 250 rpm using a Ball to Powder Ratio of 10:1. It is important to note that the initial particle sizes of aluminum powder was found to be 45 micron. However, after 2 h of milling, the particle size of powders decreases by milling process. In addition, the morphology of the aluminum powders after milling change to flaky shape. The graphene nanoplatelets distribute uniformly between flaky shape aluminum powders.

The chemical composition of the Al7068 aluminum alloy studied in this work is given in Table 1. The Al7068 aluminium alloy ingots cut into small pieces and then placed into a graphite crucible. The graphite crucible was placed in an electrical resistance. Melting of aluminium alloy was done by heating it to a temperature of ~750 °C. Then, stirring of Al alloy melt was accomplished for 10–15 min with the help of a mechanical stirrer and 60 s with 2000w ultrasonic waves to homogenise the uniform temperature throughout and adding up of pre-heated aluminum powder and graphene nanoplatelets composite with different wt% in the metal melt.

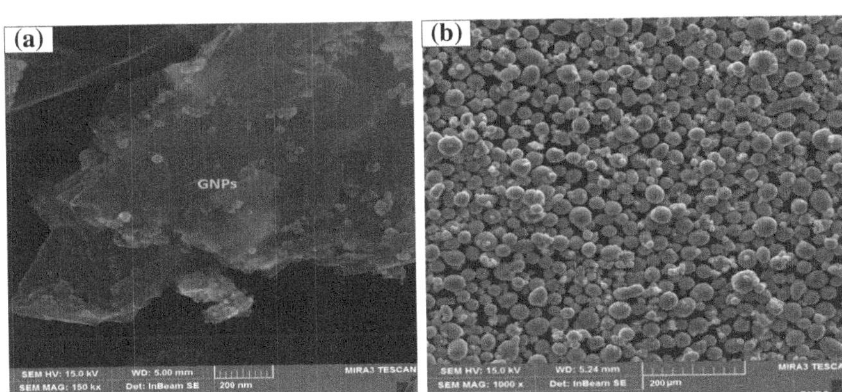

Fig. 1 SEM images of: **a** Exfoliated graphene nano plates, **b** Al powder morphology

Table 1 Chemical composition of the 7068 primary ingots (wt%)

Al	Zn	Mg	Cu	Ti	Mn	Fe	Si
Rem	8.2	3	2.46	0.1	0.1	0.16	0.03

Nanocomposites specimens have been prepared with weight% of 0.1, 0.3, 0.5, 0.7 and 1 graphene nanoplatelets reinforcements.

After successful addition of nano reinforcement particles and uniform mixing throughout, the composite melt was poured into a permanent mold prepared according to ASTM B557M-10 standard (Fig. 2). The main advantage of this mold is the application of an appropriate uphill filling system and feeding design, providing a low turbulence manner of fluid flow, which results in reduced gas entrapment and porosity in the specimens.

For structural studies, optical microscope equipped with an image analysis system (Clemex Vision Pro. Ver.3.5.025), scanning electron microscope performed in a Cam Scan MV2300 equipped with an energy dispersive X-ray analysis (EDX) accessory have been used. The cut sections were polished and then etched by Keller's reagent to reveal the structure. The average grain size of the specimens was measured according to the ASTM: E112 standard. Phase identification was also performed by X-ray diffraction method (Philips).

(a)

(b)

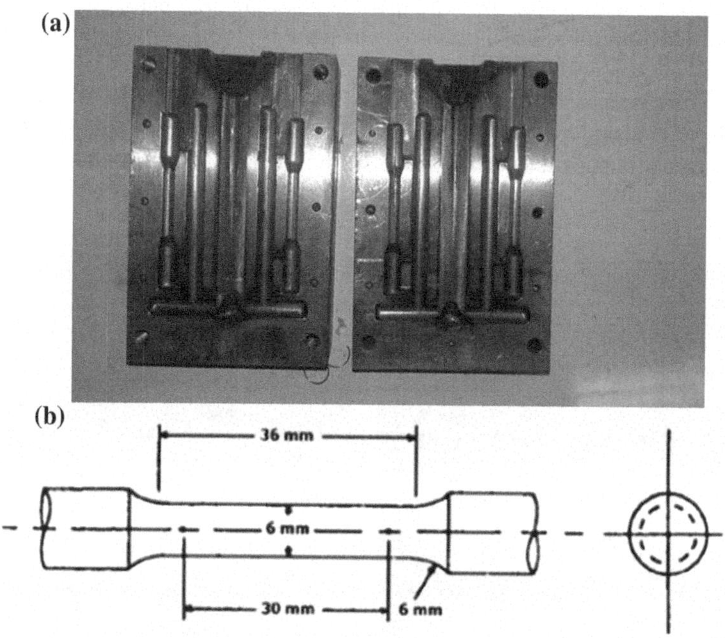

Fig. 2 Tensile specimen geometry and dimensions of: **a** cast iron mold, and **b** tensile sample dimensions

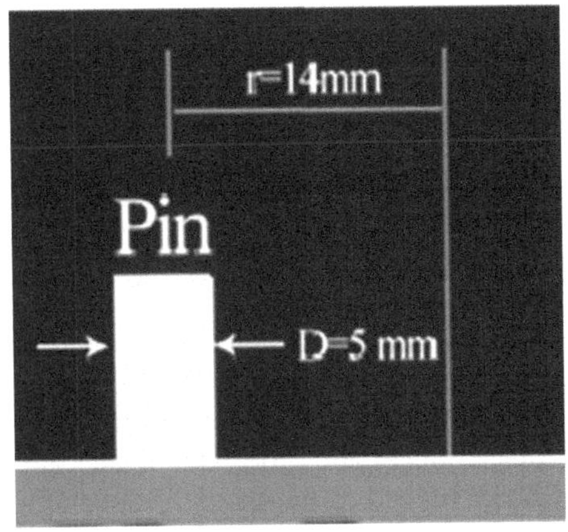

Fig. 3 Schematic of pin-on-disk configuration

Before wear testing, T6 heat treatment was applied to the castings containing optimum amount of the graphene, i.e. 0.5 wt% graphene nanoplatelet. In the T6 heat treatment, the as-cast aluminium alloy and nanocomposite samples were heated at 460 °C for 8 h and water quenched. The quenched samples were tempered at 120 °C for 24 h and were furnace cooled [2].

Dry sliding wear tests were conducted using a conventional pin-on-disc testing machine to appraise room temperature (i.e., 25 °C) wear behavior of the Al–Zn–Mg–Cu/graphene nanocomposite against a DIN 100Cr6 steel disk with a hardness of 62HRC. The pins 5 mm * 20 mm, were in a conformal contact with the steel disk. Figure 3 shows the schematic of the pin-on-disk configuration used in this study. The aluminum alloy was tested at a rotational speed of 250 rpm, corresponding to a speed of 0.5 ms^{-1}, under nominal loads of 40 and 20 N for a sliding distance of 1500 m. The weight of the specimens was measured before and after the wear test using an electronic balance (GR200-AND) with an accuracy of 0.1 mg.

Results and Discussion

Structural Characterization

It is important to note that the initial particle sizes of pure aluminum powder was found to be 45 μm (Fig. 1b). However, after 2 h of milling, the particle size of powders decreases by milling process, which can accommodate better dissolution and lower agglomeration during subsequent stir casting. In addition, the morphology of the aluminum powders after milling change to flaky shape. The graphene nanoplatelets distribute uniformly between flaky shape aluminum powders.

It is obvious that dissolution of small powders in the melt is easier relative to the larger agglomerated ones. Besides, formation of nanodispersions is mainly caused by two phenomena. During injection of powders into the melt, dissolution of aluminum powders occurs, which act as carriers of graphene particles to the melt and also protect them from any contact with the surface of the melt and alumina layer. While after dissolution, nanoparticles will be released in the matrix. Also, ultrasonic vibration prevents agglomerated particles from being sintered. It has been suggested that aluminum powders which are in the inner part of agglomerated nanoparticles act as a binder at 700 °C avoiding separation and release of nanoparticles, which only could be separated by means of ultrasonic vibration during stirring.

Figure 4 shows the effect of various amounts of graphene nanoplatelet on the average grain diameter of the specimens. The study of specimens showed the presence of different microstructural features, which may result in different mechanical properties. From Fig. 4, the optimum amounts of graphene nanoplatelet was determined to be 0.5 wt%. Several mechanisms have been proposed for the grain refining process. In some mechanisms the presence of some particle is known to be effective for grain refinement procedure. Graphene nanoplatelet is a potent nucleating site for aluminum. 7068 aluminum cast alloy without any graphene nanoplatelet addition have the maximum grain size (550 μm). The present experimental results confirmed that, the addition of graphene nanoplatelet to 7068 aluminum alloy significantly refines the coarse columnar primary α-Al grains to fine equiaxed α-Al grains (70 μm) due to the presence of graphene nanoplatelet which are nucleating agents during the solidification of α-Al grains.

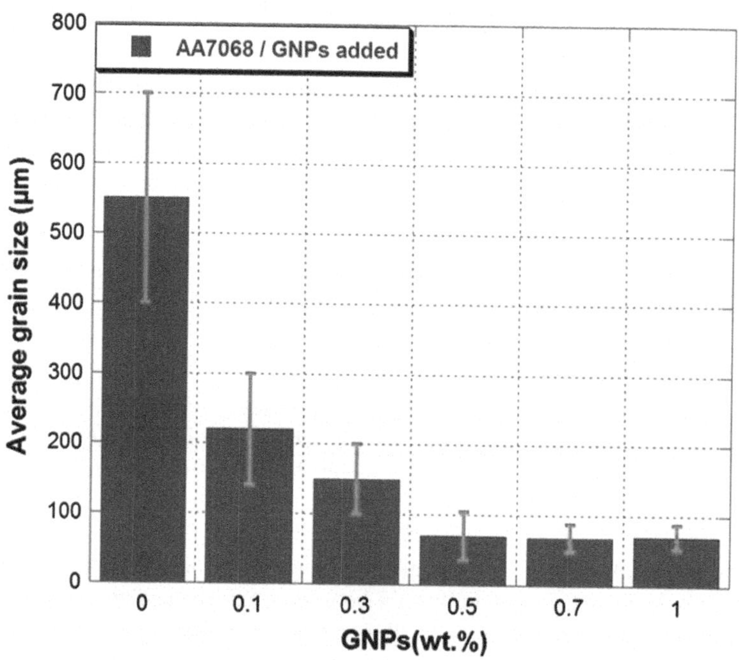

Fig. 4 Grain size of with various graphene nano plates contents

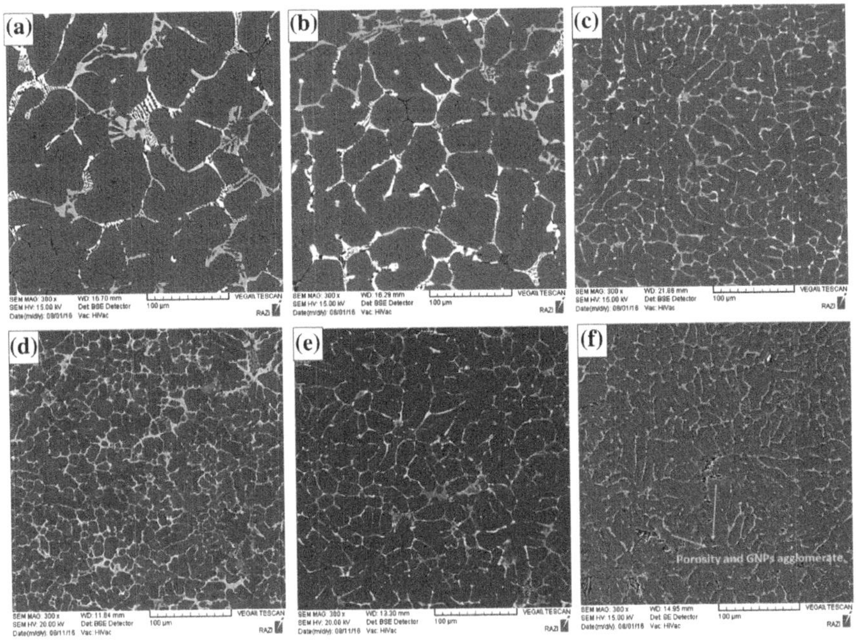

Fig. 5 SEM images of microstructures of AA7068, with **a** 0.0 wt%, **b** 0.1 wt%, **c** 0.3 wt%, **d** 0.5 wt%, **e** 0.7 wt% and **f** 1 wt% graphene nano plates

The microstructures of the 7068 aluminum alloy cut from castings after adding graphene with different percentage are shown in Fig. 5. This image shows the change in dendrite morphology of the 7068 aluminum alloy after adding graphene. The microstructures of alloy revealed a rosette-like microstructure of primary α-Al grains solid solution surrounded by interdendritic secondary phases. In comparison with graphene added specimens, unrefined specimens showed coarser morphology. From Fig. 5, it is noticeable that graphene and stir casting enhances the number of grain boundaries and therefore promotes a more homogeneous distribution of intermetallic precipitates. The optimum content of graphene nanoplates according grain size and wear behavior is 0.5 wt%.

The lamellar eutectic structures can be observed in Fig. 6. From Fig. 6, it is clear that the predominant eutectic structure includes α-Al and η-$MgZn_2$ phases. It is important to note that the solubility of η-phase for copper is relatively high, so it can be observed in the eutectic structure. Figure 6 shows the distribution of the major elements by scanning electron microscopy and local analysis in sample S0. Analysis of the EDX result of $MgZn_2$ indicated that copper and aluminum contents are higher than the average level of chemical composition in the alloy. In spite of more contribution of aluminum, matrix may be involved in the EDX signal of the $MgZn_2$ because of its small size. It is obvious that the $MgZn_2$ dissolved the Cu and

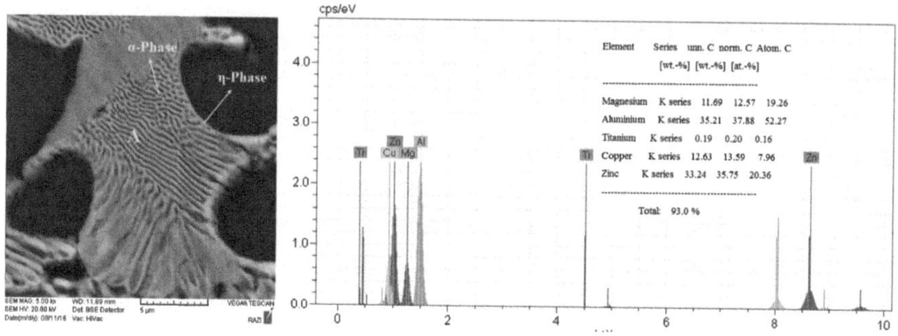

Fig. 6 Microstructures and EDX analysis of eutectic phase in as-cast AA7068 specimen

Al elements, and formed the Mg (Zn, Cu, $Al)_2$ phase, which is consistent with the present literatures [5].

The segregation of solute that occurred during casting led to the high concentration of Cu, Mg and Zn in the interdendritic eutectic regions. The diffusion velocity of copper is lower than zinc and magnesium, which resulted in the higher concentration of copper in the regions of eutectic structures during solidification. The driving force for the phase transformation from Mg (Zn, Cu, $Al)_2$ to Al_2CuMg must be super saturation of copper in the regions of eutectic structures, which makes Al_2CuMg a stable phase.

Solidification Mechanisms

Three class of models has been developed regarding incorporation of particles into a solidifying matrix, namely (i) kinetic models predicting velocity of the solid/liquid interface which is critical for the transition from particle pushing to engulfment, (ii) thermodynamic models [10] corresponding to classical heterogeneous nucleation theory, (iii) models based on the ratio of the thermophysical properties of the particles and the melt [11, 12]. The latter predicts the incorporation ability in general that means dependency of incorporation behaviour on the processing conditions. Accordingly, any dependence on the morphology of the interface is not included.

These two models demonstrate that thermal conductivity of particles incorporated into the liquid matrix is directly related to possibility of particle through the grains of the solidifying matrix due to change of the interface shape from convex to concave [12–14]. In fact, temperature gradient ahead of the solidification front is affected by thermal conductivity of the particles. Therefore, lowering thermal conductivity of the particles acts as a barrier to the removal of the heat necessary for further solidification and consequently inhibits particle.

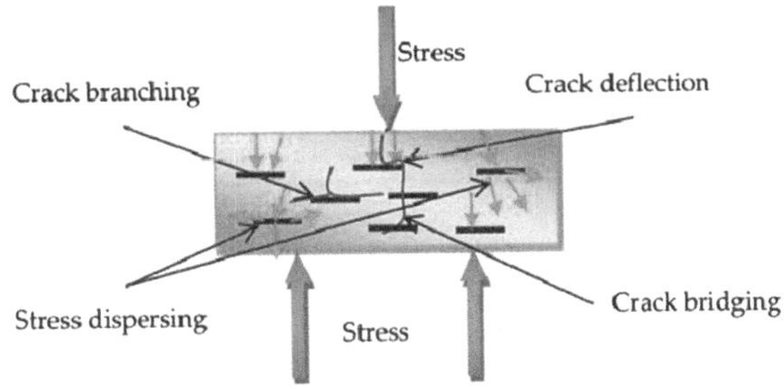

Fig. 7 Scheme of AA7068–graphene nano plates nanocomposite under load

In the present study, thermal conductivity of graphene nanoplatelet is expected to be very large, which is related to high thermal conductivity of graphene nanoplates [15–17]. This could be explained by thermal conductivity of graphene nanoplatelet being better conserved in graphene nanoplatelet than in single layer ones.

Some studies found that the high improvement in the strength of GNP composites was due to various toughening mechanisms, including stress dispersing, crack deflection, crack bridging, and cracking branching (Fig. 7) [18, 19]. The strong bondability can increase the load-transfer efficiency and eliminate the destruction of stress concentration. In addition, the GNP particles provide a higher resistance to crack propagation. For the tensile test, the first crack form at the sample. When the crack reaches the surface of graphene nanoplatelet, the crack develops along the interface between the graphene nanoplatelet and aluminum matrix. Thus, the presence of graphene nanoplatelet causes crack branching or deflection. The crack branching or deflection mechanism could increase the path of crack development, which improves the mechanical strength of aluminum matrix. Moreover, the crack bridging mechanism of graphene nanoplatelet could absorb the more energy effectively, when the composite is under external load. The whole mechanical map is presented in Fig. 7. In a word, due to the high tensile strength, high Young's modulus, and unique two-dimension morphology, Graphene nanoplatelet could reduce the stress concentration and prevent the development of the cracks, thus enhancing the mechanical strength of the aluminum matrix. These results are helpful in understanding the reinforcing effect of nanofillers used for metal-based materials.

Wear Behavior of AA7068/Graphene Nano Plate Nanocomposite

Figure 8 shows the variation of coefficient of friction with normal load for 7068 aluminum alloy for unrefined and nanocomposite specimens with and without T6 heat treatment at constant sliding speed of 250 rpm. The results indicate that the coefficient of friction decreases with increasing normal loads in all cases. However,

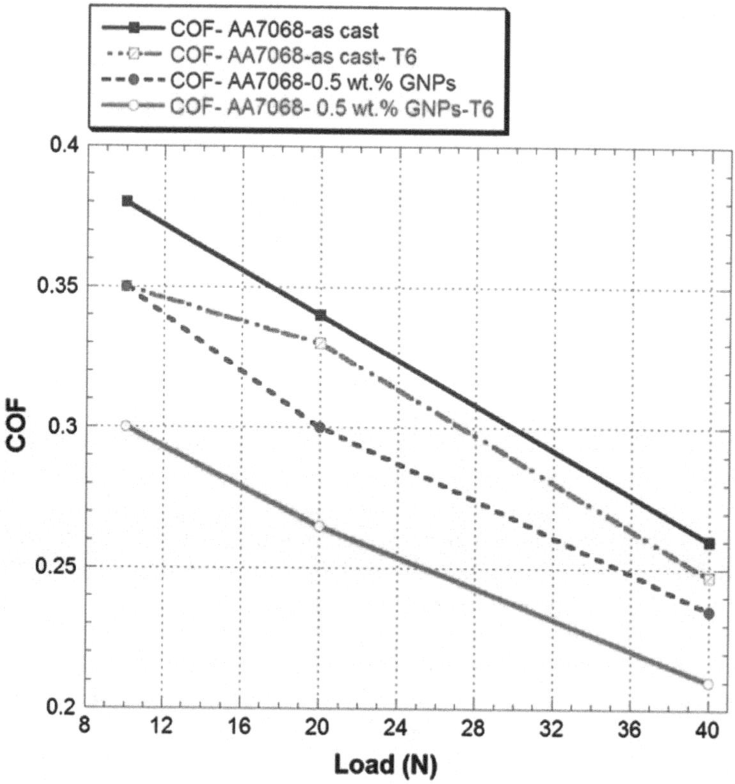

Fig. 8 Coefficient of friction values of the nanocomposite as function of load amount

the rate of decrease in coefficient of friction with normal load is significant at higher normal loads when compared to lower normal loads. The results show that the coefficient of friction change significantly by adding 0.5 wt% of graphene nanoplatelet to the aluminum matrix because of sufficient amount of solid lubricant available at the contact surface. However, higher weight percentage of graphene nanoplatelet (0.5 wt%) decreased the coefficient of friction of the composite sample significantly in comparison with other samples.

The wear results for 7068 aluminum alloy for unrefined and nanocomposite specimens with and without T6 heat treatment is shown in Fig. 9. Figure 9 shows the amount of weight loss against sliding distance. This diagram was obtained at constant normal loads (20 and 40 N) and a constant rotation speed of the counter disk (250 rpm). It can be seen that the amount of weight loss has increased by increase in sliding distance. Furthermore, increase in the amount of weight loss with sliding distance, approximately has a linear trend. Comparison between unrefined and nanocomposite specimens with and without T6 heat treatment for 7068 aluminum alloys is shown in Fig. 9.

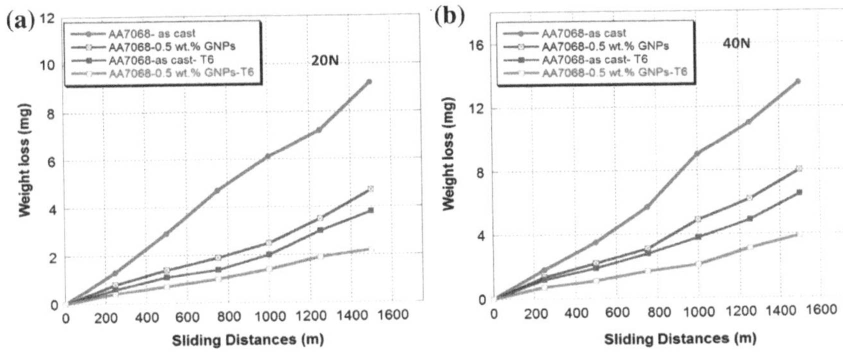

Fig. 9 Weight loss as a function of sliding distance for unrefined and refined nanocomposites with and without T6 heat treatment, **a** 20 N and **b** 40 N

From Fig. 9, it is clear that the addition of graphene nano plate to 7068 aluminum alloy has reduced the weight loss in comparison with unrefined aluminum alloy. This can be described in terms of uniform distribution and dispersion of graphene nano plate in refined by 0.5 wt% graphene nanoplatelet. It is interesting to note that, the weight loss of the T6-tempered alloy, which was generally lower than those of untreated alloys in dry sliding wear tests. Probably due to the reduced grain size of the castings, leading to a finer distribution of second phases in AA7068–0.5 wt% graphene nanoplatelet–T6 nanocomposite. Wear surfaces of unrefined and nanocomposite specimens by 0.5 graphene nanoplatelet with and without T6 heat treatment are shown in Fig. 10. It can be seen a mild wear regime grooves the wear mechanism of the all alloys in Fig. 10.

It clearly shows plastic deformation and delamination characteristics. Under an applied load of 40 N, two different regions are distinguished on the worn surface: cavities or craters and smooth regions marked as A and B in Fig. 10, respectively. In the smooth regions, the surface disclosed fine grooves and ploughing which suggest that abrasive wear was prevailing mechanism in these regions. Detached layer in the form of craters or cavities indicates locally adhesive wear due to the formation of and breaking of micro-welds during sliding. In addition, the wear rate of refined alloy by 0.5% graphene nanoplatelet with T6 heat treatment is lower than unrefined in all applied loads. It seems that during the early stages of sliding, delamination wear is the dominant mechanism for the unrefined and nanocomposite specimens by 0.5% graphene nanoplatelet with and without T6 heat treatment. For unrefined 7068 aluminum alloys with coarse dendritic morphology can be easily broken and contribute to the weight loss.

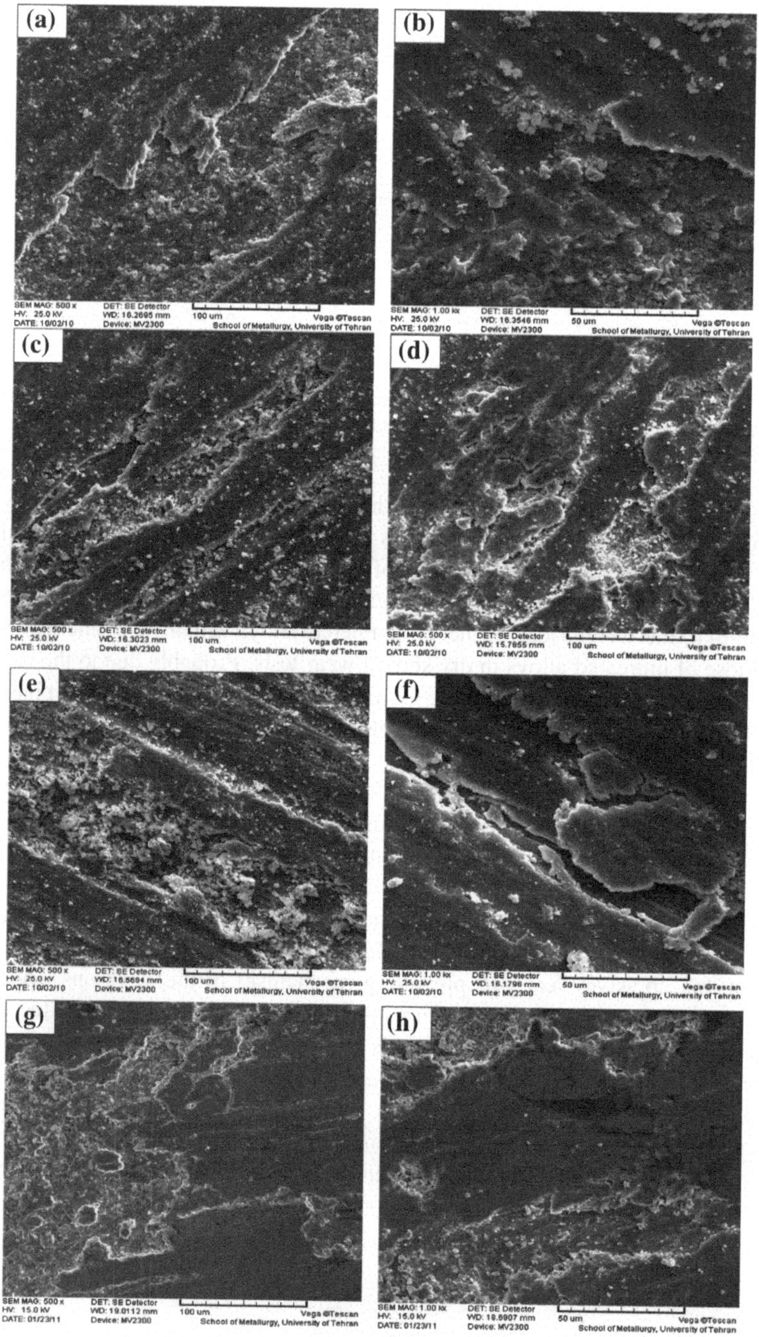

Fig. 10 Wear surface of the AA7068 after 1500 m distance with 40 N load during wear process of: **a** and **b** unrefined, **c** and **d** refined by 0.1 wt%, **e** and **f** refined by 0.5 wt%, **g** and **h** refined by 0.5 wt% graphene nano plates after T6 heat treatment

Conclusions

In this study, 7068 aluminum matrix nanocomposites reinforced by graphene nanoplatelet was synthesized and the microstructural and tribological behavior of these nanocomposites was investigated. The experimental findings are summarized as follows:

1. Stir casting method with ultrasonic waves is very effective for improvement in mechanical properties of Al7068 aluminum alloy matrix graphene nanoplatelet reinforced composites and achieves uniform distribution of graphene nanoplatelet in the aluminum matrix.
2. The optimum amount of nanoparticles is 0.5 wt% graphene nanoplatelet.
3. When the wt% graphene nanoplatelet reaches more than 0.5 wt%; agglomeration of graphene nanoplatelet at the grain boundaries causes embrittlement, porosities, less interfacial bonding and so decrease in mechanical properties.
4. The coefficient of friction of the 7068 aluminum alloy and nanocomposite reinforced by GNP decreases with increasing the normal load.
5. Increasing amount of graphene nanoplatelet to 0.5 wt% significantly improved the COF of the 7068-aluminum alloy.
6. The wear rate of the 7068-aluminum alloy and nanocomposite reinforced by graphene nanoplatelet increased with increasing the normal load.
7. The wear rate of the 7068-aluminum alloy and nanocomposite reinforced by graphene nanoplatelet decreased with T6 heat treatment.
8. The SEM investigation of the worn surfaces had shown that abrasive wear was the main wear mechanism in these composites.
9. The lowest wear rate and coefficient of friction of the AA7068–0.5 wt% graphene nanoplatelet is attributed to self-lubricating behavior of the nanocomposites.

References

1. Haghparast A, Nourimotlagh M, Alipour M (2012) Effect of the strain-induced melt activation (SIMA) process on the tensile properties of a new developed super high strength aluminum alloy modified by Al–5Ti–1B grain refiner. Mater Charact 71:6–18
2. Alipour M, Emamy M (2011) Effects of Al–5Ti–1B on the structure and hardness of a super high strength aluminum alloy produced by strain-induced melt activation process. Mater Des 32:4485–4492
3. Alipour M, Emamy M, Farsani RE, Siadati MH, Khorsand H (2015) Effects of a modified SIMA process on the structure, hardness and mechanical properties of Al–12Zn–3Mg–2.5Cu alloy. Iran J Mater Sci Eng 12:77–88
4. Alipour M, Aghdam BG, Rahnoma HE, Emamy M (2013) Investigation of the effect of Al–5Ti–1B grain refiner on dry sliding wear behavior of an Al–Zn–Mg–Cu alloy formed by strain-induced melt activation process. Mater Des 46:766–775

5. Alipour M, Emamy M, Ebrahimi SHS, Azarbarmas M, Karamouz M, Rassizadehghani J (2011) Effects of pre-deformation and heat treatment conditions in the SIMA process on properties of an Al–Zn–Mg–Cu alloy modified by Al–8B grain refiner. Mater Sci Eng: A 528:4482–4490
6. Alipour M, Emamy M, Azarbarmas M, Karamouz M (2010) Effects of Al–5Ti–1B master alloy on the microstructural evaluation of a highly alloyed aluminum alloy produced by SIMA process. In: AIP Conference Proceedings, vol 1252, pp 1060–1072
7. Alipour M, Emamy M, Rasizadeh J, Karamouz M, Azarbarmas M (2011) Effects of Al–8B grain refiner on the structure, hardness and tensile properties of a new developed super high strength aluminum alloy. TMS Annual Meeting, vol 2, pp 309–320
8. Pradeep Kumar GS, Koppad PG, Keshavamurthy R, Alipour M (2017) Microstructure and mechanical behaviour of in situ fabricated AA6061–TiC metal matrix composites. Arch Civ Mech Eng 17:535–544
9. Alipour M, Emamy M, Rasizadeh J, Karamouz M, Azarbarmas M (2011) Effects of Al–5Ti–1B grain refiner on the structure, hardness and tensile properties of a new developed super high strength aluminum alloy. TMS Annual Meeting, vol 3, pp 833–842
10. Alipour M, Azarbarmas M, Heydari F, Hoghoughi M, Alidoost M, Emamy M (2012) The effect of Al–8B grain refiner and heat treatment conditions on the microstructure, mechanical properties and dry sliding wear behavior of an Al–12Zn–3Mg–2.5Cu aluminum alloy. Mater Des 38:64–73
11. Mirjavadi SS, Alipour M, Hamouda AMS, Besharati Givi MK, Emamy M (2014) Investigation of the effect of Al–8B master alloy and strain-induced melt activation process on dry sliding wear behavior of an Al–Zn–Mg–Cu alloy. Mater Des 53:308–316
12. Afshari BM, Mirjavadi SS, Dolatabad YA, Aghajani M, Givi MKB, Alipour M, Emamy M (2016) Effects of pre-deformation on microstructure and tensile properties of Al—Zn—Mg—Cu alloy produced by modified strain induced melt activation. Trans Nonferrous Met Soc China (English Edition) 26(9):2283–2295
13. Alipour M, Mirjavadi S, Besharati Givi MK, Razmi H, Emamy M, Rassizadehghani J (2012) Effects of Al–5Ti–1B master alloy and heat treatment on the microstructure and dry sliding wear behavior of an Al–12Zn–3Mg–2.5Cu alloy. Iran J Mater Sci Eng 9(4):8–16
14. Alipour M, Emamy M, Rasizadeh J, Azarbarmas M, Karamouz M (2011) Effect of predeformation and heat treatment conditions in the modified SIMA process on microstructural of a new developed super high-strength aluminum alloy modified by A1–8B grain refiner. TMS Annual Meeting, vol 3, pp 843–853
15. Alipour M, Emamy M, Rasizadeh J, Karamouz M, Azarbarmas M (2011) Effects of Al–8B grain refiner on the structure, hardness and tensile properties of a new developed super high strength aluminum alloy. TMS Annual Meeting, vol 2, pp 309–320
16. Binesh B, Aghaie-Khafri M (2016) RUE-based semi-solid processing: Microstructure evolution and effective parameters. Mater Des 95:268–286
17. Binesh B, Aghaie-Khafri M (2015) Microstructure and texture characterization of 7075 Al alloy during the SIMA process. Mater Charact 106:390–403
18. Ranjbar N, Mehrali M, Mehrali M, Alengaram UJ, Jumaat MZA (2015) Comprehensive study of the polypropylene fiber reinforced fly ash based geopolymer. Cem Concr Res 76:222
19. Baradaran S, Moghaddam E, Basirun WJ, Mehrali M, Sookhakian M, Hamdi M, Nakhaei Moghaddam MR, Alias Y (2014) Mechanical properties and biomedical applications of a nanotube hydroxyapatite-reduced graphene oxide composite. Carbon 69:32

Author Index

A
AbdelMeguid, Mohamed E., 119
Abuzin, Yu. A., 233
Adhikari, Saikat, 59
Alipour, Mohammad, 233
Arindam, Paul, 203
Arvind Singh, R., 135

B
Bahei-El-Din, Yehia A., 119
Bharadwaj, Shikhar, 193
Bojarevics, A., 183
Bowen, P., 87

C
Cadet, Stevens, 221
Chen, Fei, 149

D
Dear, M., 87
Doel, T. J. A., 87
Dussing, Matthew, 71

E
Elsharkawi, Mohammed, 29
Esawi, Amal M.K., 29

F
Farsani, Reza Eslami, 233

G
Gerbeth, G., 183
Giri, Anirban, 59
Grants, I., 183
Gupta, Manoj, 193
Gupta, Sheetal, 59

H
Harrigan, William Jr., 19
Harrigan, William C., 3
Hasegawa, Yoshiko, 103
Hatem, Tarek M., 119
Hattori, Koutarou, 103
Hegazi, Hesham A., 119
Hirami, Shogen, 103

I
Iqbal, Khurram, 221
Izui, Hiroshi, 103

J
Jayalakshmi, S., 135
Jin, Xueze, 43

K
Komiya, Yoshiki, 103

L
Lavernia, Enrique J., 71, 149
Lin, Yaojun, 149

M
Ma, Kaka, 71
Manigandan, K., 149

P
Parker, Andrew, 19

R
Reactivity, 221

S
Sarma, M., 183
Schoenung, Julie M., 71
Shalan, Khalid M., 119
Shan, Debin, 43
Sommer, Al, 19
Sommer, Mark, 19

T. S. Srivatsan et al. (eds.), *Metal-Matrix Composites Innovations, Advances and Applications*, The Minerals, Metals & Materials Series,
https://doi.org/10.1007/978-3-319-72853-7

Srivastava, Vivek, 59
Srivatsan, T.S., 135, 149, 193, 203
Stanley, H., 87

T
Tekumalla, Sravya, 193
Topping, Troy D., 71

V
Verduzco, Miguel, 19

X
Xu, Wenchen, 43

Y
Yang, Hanry, 71

Z
Zeng, Xiangqian, 43
Zhang, Yuzheng, 19

Subject Index

A
$Al_{18}B_4O_{33}$ whisker, 44, 53, 57
Al alloy matrix composites, 72, 78, 82, 84
Alumina fiber, 138, 139, 144
Aluminum Matrix Composites, 3–8, 10, 12–17, 43, 44
Aluminum matrix nanocomposites (AMnCs), 19–21
Archimedes' method, 107

B
Bimorph PFC (PFCB), 125

C
Carbon nanotubes, 30, 31
Casting, 30, 31, 33, 34, 37, 38
Cavitation treatment, 184–187, 189–191
C/Cu composites, 221, 222, 227
Coefficient of thermal expansion, 195, 198, 201
Colloidal tin-palladium activation, 31–33, 35, 38
Consistency, 64
Corrosion resistance, 26
Crack arrest, 90, 91, 93–98, 100–102
Cu-Nb microcomposite, 205–208, 210, 212, 214, 216–218

D
Damage modelling, 125
Delamination, 125, 127
Density, 195, 198, 200
Discontinuously-reinforced, 151–154, 156, 177
Ductility, 194, 200, 201

E
Electroless copper-cobalt plating, 30–33, 35, 36, 38
Environmental effects, 88, 89, 95, 96
Extrusion, 194, 196, 197, 199, 201

F
Fatigue crack growth, 88–94, 96–101
Fatigue resistance, 25
Fibre pull out, 89, 96, 99, 102
Friction, 136, 137, 140, 145–147

G
Gas-Liquid Interactions, 162

H
Hardness, 136–142, 146
Homogeneity, 64
Hot compression, 45, 46, 50, 53, 54, 57
Hubble Space Telescope, 6

I
In-situ, 60, 63

K
Kinetic analysis, 47

L
Liquid-Liquid Interaction, 165

M
Macro-hardness, 65, 66, 68
Macroscopic morphology, 48, 50, 54
Magnesium alloy nanocomposite, 194
Magnesium alloys, 136–138, 140–147
Magnesium composites, 146
Manufacturing, 7, 15
Matrix microstructure, 137, 138, 140, 141
Mechanical behavior, 80, 84
Mechanical properties, 44, 55, 57, 60, 62, 63, 65, 68, 234, 238
Metal-matrix composites (MMCs), 60, 150–154, 160, 161, 164, 168, 172–175, 177
Metal matrix nano composites (MMNCs), 234

T. S. Srivatsan et al. (eds.), *Metal-Matrix Composites Innovations, Advances and Applications*, The Minerals, Metals & Materials Series,
https://doi.org/10.1007/978-3-319-72853-7

Micro-hardness, 195, 199, 207–209, 214, 218
Microstructural effects, 205, 206, 214
Microstructure, 44, 45, 50, 53, 54, 56, 62, 63, 75, 172, 195, 196, 201, 234, 235, 239, 240
MMC production, 184
Modal Analysis, 121, 126, 129
Modulus, 59, 65–68

O

Orthogonal array, 104, 108–111

P

Palladium acceleration, 31, 33
Particle Reinforced Aluminum, 3, 8, 10, 12, 17
Piezoelectric Fiber Composites, 120, 121, 128
Piezoelectric response, 121
Pneumatic powder injection, 61, 64
Porosity, 195, 198
Powder processing, 235, 237, 238
Power dissipation, 50–52
Processing maps, 44, 50, 52–54
Processing techniques, 150–154, 162, 165, 166, 177
Processing variables, 60, 149

S

Solidification Mechanisms, 240
Space Shuttle Orbiter, 4
Spark plasma sintering, 106
Spray atomization and deposition, 151, 152, 160–164, 166, 167, 177
Steel composites, 184
Strength, 194, 199–201
Strengthening mechanism, 21, 81
Structural Health Monitoring, 120, 121, 128, 129
Surface modification, 205, 206, 217, 218

T

Taguchi methods, 104, 109, 117
Temperature, 88–91, 93–102
Tensile fracture, 212, 214–218
Tensile properties, 23, 62, 66
Tensile response, 205, 210
Third body, 144, 146
TiC, 60, 61, 63–68
Titanium matrix composites, 104, 108, 112, 116
Titanium metal matrix composites, 88
Transformation Field Analysis, 121, 122, 128

U

Ultrafine grained structure, 72–79, 81–84
Ultrasonic waves, 234, 235, 238

W

Wear, 136, 137, 140–147
Wear behavior, 104
Wear resistance, 22
Wetting, 221, 222, 224–227

X

X-Ray Diffraction, 64, 198

Y

Yttria-stabilized zirconia, 72, 73

The manufacturer's authorised representative in the EU is Springer Nature Customer Service Centre GmbH, Europaplatz 3, 69115 Heidelberg, Germany. If you have any concerns regarding our products, please contact ProductSafety@springernature.com

Printed and bound by CPI Group (UK) Ltd, Croydon, CR0 4YY
23/09/2026
02221528-0008